Mental Physics

By Nathaniel Durham

Copyright© 2011

All Rights Reserved

This book goes to David Paul Slepitis.

Introduction

My name is Nate Durham and this is a joke for an introduction.
I am not sure of this and this is pure
Speculation, but in previous forms from
The little paramecium creature to Higher lifeforms,
I might have had Previous incarnations, but none like
This. In my possible previous forms, for example I had no hygienic habits or
any way to acquire them. As I died and assumed higher and higher lifeforms, I may have been
eaten or squished many times but as my incarnations progressed I may have acquired some
hygienic
instincts that consisted of eating and licking away of bugs (including
parasites) and dirt as well as various wounds. Suddenly, I seemed to come
into this modern world. How different and demanding it was! I had to adopt
much more complex and much more sanitary survival habits and conform to strange and more
strict societal standards! I was in a new setting that
we all know as Civilization! This world of civilization as we know it
brought to this world many of my inspirations and I was inspired to try
to figure out the answers to some of the questions encircling existence. As a result, I wrote this
book!
REMEMBER, THIS IS JUST PURE SPECULATION
OF WHERE AND WHAT I CAME FROM!
ENJOY THIS BOOK!

The Nature Of Field Interaction

If space-time can be considered an extension of matter by the means of a
field between more than one mass, then if the extension of the mass

relative to the other was divided by The mass itself, then we could have
relative space.

If the same applies for another mass, then if the two masses are multiplied by each other and
then divided by the sum of the strengths
of the fields, then we may get our true space and this could produce
our space curvature. The curvature of space theoretically produces gravitational force
according to Albert Einstein.
Electric and magnetic fields act at a shorter distance and this may be a
result of Electromagnetic force more
nearly equaling the mass and are therefore being more mass restricted.
Strong nuclear force is probably the most confined by the
same possibility of nearly equaling energy equivalent of the mass.

The attraction of opposite electric or
magnetic charges may result from an "observation of each others"
opposite state of electromagnetic existence. On the other hand, two parallels may not observe
each other as they are doing the same
thing and will produce a crunching of space-time that will be opposed.
Weak nuclear force may act in the same way as too, many protons or
neutrons will produce a similar state of existence which may produce this
weak force. Since everything is relative, then like charges will repel until they become relative
and this may result from a paradox in which they are faster than each other which may finally
reach a same speed by repelling each other until such a standstill begins. It is arguable that they
will just keep moving through space, but eventually, gravity will pull it back
and an equilibrium may be established.

Also, if negative and positive charges are attracted to a certain point, then they may eventually
reach a state that produces a like reality which may cause them to be held away but not repelled
unless forced
to move farther. However, as they are repelled, they may begin to form
opposite realities once again and may decelerate, stopping at the
point where the attraction boundary is and this may explain atoms.
Electromagnetism may be explained by the rate of change of an electric
field from one given relative point to another. At the point at which a given
position is reached by a electrically charged particle, the electric field has hit its peak and the
greatest amount of magnetism is produced. At a very tiny quantum level, the magnetic field
strength is just as high on the opposite point, but is also the opposite charge because the change
is in the opposite direction. Time could be taken into account at this point, as time is one of the
factors that determines this
change in the strength of the
magnetic field and this determines

the speed at which the particle is moving.
If you move an object that has an
electric or magnetic charge back and
forth, it will produce changes in the
magnetic field that will be
propagated over time and the
propagation through time will be
exactly equal to the propagation
through space and therefore, space
and time may be unified.
On the other hand, since everything is relative,
then an uncharged object moving
through the electric or magnetic field
of a charged one may also produce
electromagnetic waves that may be
inverse in position relative to the
motion of the uncharged particle as
well as cause this uncharged particle
to temporarily acquire an
electromagnetic charge.

Finally, to make one note on gravitational force,
the shape that matter may give the
space-time around it may be
dependent on the distribution of
graviton particles and therefore, the
shape of space may be determined
by the shape that the distribution of
graviton particles takes.

Also, space may consist entirely of graviton
particles, since the trough of a
gravitational wave may, due to the
sine geometry of the wave, represent
the collapse of gravity instead of
space, which could produce the
space, since the crest and trough of
the wave are only a matter of
perspective. Therefore, as the
gravitational waves get farther away
from the source then they could

increase in wavelength and the space
could become closer to being flat.
The result could be gravitational
waves that, between two masses,
form a wavelength equal to the
difference between the two or more
wavelengths if the masses were
alone and this could be expressed as
gravitational space curvature. The
consequence may be that space-time
is highly manipulable.

Also, if we were to view a space that
was produced by a stronger force
such as electromagnetic or weak,
then we could possibly
look into a smaller, denser
pace and a more gravitational
universe could be much larger.
Another possibility for the reason
that stronger forces act at shorter
distances is that perhaps the
stronger forces produce the same
amount of energy over the same
period of time, but because of this,
the stronger forces are not able to
consist of as many virtual particles
and that because of this, the field
strength reaches zero particles per
cubic meter or other cubic unit in a
shorter amount of time and at a
shorter distance, but if a particle of
strong nuclear force having a
specific frequency collides with a
photomultiplier tube, which is used
to measure or detect electromagnetic
radiation, then the frequency might
register as being 100,000 times as
high, since strong nuclear force is
100,000 times as powerful as
electromagnetic force.

Also, in a universe that uses stronger forces
than gravity to compose relative
space and time, bodies of matter may

pass through time more quickly, due
to the curvature of that space-time
which would cause events to occur
more closely together due to the
looping of space-time that would
allow events to occur more closely
together.

An example would be that if
a person was within and influenced
by an electromagnetic field then the
gravitational space-time would be
bent by the electromagnetic field as a
result of the differences in their
ability to curve space-time and
everything outside of that field would
appear to occur more rapidly to that
person than if that person was not
within that electrical field.
The conclusion that I come to is that
weaker forces can be bent by
stronger forces and that the relativity
of space-time is dependent on the
strength of the field that the particle
or body is influenced by. Also I am
theorizing that the differences in the
relativity of various forces may cause
space-time to loop or coil up.

The final statement of this theory is that
space-time may be completely
composed of the forces that warp it.
If any of these forces are polar, then
if the polarity of these forces is
reversed rapidly enough, then time
travel may be possible by the
reversal of nearby space-time, but this could produce waves of
immense force and the consequences may possibly be
catastrophic to surrounding bodies of
matter.

The Expanding Universe And Departing Galaxies

Although though the Hubble
telescope findings show that the galaxies are spreading, I find
it possible that there may be a
gravitational influence behind this.

I speculate that there might
be immense numbers of galaxies and
other clusters of matter outside of the
distance at which the Hubble can
detect and this large mass may be
pulling the galaxies toward it, that is
the enormous mass of those other
galaxies.

Once everything is pulled
into these areas of space, then these
galaxies may merge to form larger
galaxies and then the same effect may
take place, which is that these new
galaxies may be pulled farther apart,
but yet merge with some other super
massive galaxies.
However, if these
galaxies become large enough, they
may gravitationally implode and
collapse into black holes. This process
may occur over and over again until
the universe becomes a
singularity again.

Galaxies might also only reach a specific size before
the space-time within and around them
gets saturated enough with

gravitational waves, which consist,
theoretically of graviton particles
that the galaxy gravitationally
collapses.

On the other hand, this
may serve as a control to prevent the
universe from expanding too, fast by
gravitationally curving the
edge of the universe just as the
insides of the universe may be
pulled apart by a gravitational "Tug-Of-
War." The next possibility
may be that when a "Tug-Of-War"
within the universe becomes
strong enough, then there may be a
distortion of space-time
between the bodies that are fighting
over which one gets that
piece of matter, which may teleport
this piece of matter to the
edge of the universe.
As I said earlier
in a different phrasing, the
gravitational "Tug-Of-War" and the
gravitational curvature of
space-time at the edge of the
universe may produce the
equilibrium that the universe
requires in order to keep itself
in one piece.

Another possibility is that since the
void lacked space-time that the void
collapsed upon itself to form the
singularity that produced the
universe. As the tension of the
singularity built up, it equaled the
force of the collapse of the
singularity and then the equaled
collapsing force produced a wave
that sent out the signal to the
remaining void that told it that it

could not collapse any farther and
this may have caused the
compressed void to expand as the
wave served the purpose of opening
up more of the void which may have
helped the universe accelerate its
expansion and this wave may be
attached to the edge of the universe
which on contact with the void may
relieve the tension of the void.
When the tension of the void us released,
this may produce some explosions of
energy that when used up may form
matter and as the universe gets bigger,
the wave or "ripple" surrounding
it may cover more area
with every movement, accelerating the
expansion as long as the wave travels.
Because the traveling wave has no
space-time to move through as a
medium, this may provide some
possibility of the expansion moving at
a higher speed than light. The faster
movement than light by this expansion
is currently a bit of a hot scientific
discussion.

The concept of energy

Another force may be involved in the
accelerated expansion of the
universe. Albert Einstein defined
matter as a concentrated form of
energy.

When matter is converted to
its equivalent, which is the potential to
cause disturbance or in other words,

its energy, the energy becomes infinite
and so does its mass.

At this point, there is no difference between matter
and energy and this may only occur at
the speed of light, which is 186,424
miles per second.
When matter is not moving at the
speed of light, but some of it is
converted to its equivalent in energy,
this matter becomes pure kinetic
energy and speeds away at the speed
of light and as a result, becoming light.
Furthermore, inertia, which is defined
in textbooks as the resistance to
change in motion is equal to and is the
amount of work required to change the energy of one
mass back to its original state of energy and as such
occurs, the acting force will either lose nor
acquire the same amount of energy
that it changed in one of two forms potential energy or kinetic
energy.

Potential energy is energy that can be but
hasn't been converted, which in textbook format is the
lack of motion while kinetic energy in the same
textbook format is motion. However, the kinetic energy
that alters the motion of another mass by converting
potential energy could be considered the potential energy
instead of the kinetic,

Since the kinetic hasn't done the
work yet and the potential energy can be considered the
kinetic energy because it is doing work by altering the
kinetic energy of the moving mass. In either form, energy
is also a relative force and are of equal nature. A slight
mass can become infinite energy at the speed of light just
as energy can be infinitely concentrated
when a mass is perfectly stationary and this seems to
point perfectly toward Albert Einstein's famous equation
"E=MC^2."

Parallel Universes, The Expansion of the universe And Gravitational Force

The expansion of the universe may
have another cause. Their, of course
may be parallel universes and these
universes may interweave their space
times in such a manner that one
universe's space may be woven with
the which may cause a mutual
overlapping of all other universes'
space-times which could underlie the
curvature of space-time that is
regularly discussed by many
physicists.

This gravitational folding
may be the glue of all universes as
well as form a portal to all other
universes, although we probably
would not observe our transfer. This
overlapping of all universes space-time
may indicate that all matter which
we are gravitationally attracted to,
which is all matter including the little
toys all over the bathroom as well as
the bathtub may be directly tied to
another universe.

Such hyper dimensional gravitational
force may only be observed in a single
universe as typical curved or warped
space-time, but looks may be

deceiving!

Another possibility for other
universes to exist is that the number of
other universes could be equal to the
square of the number of relations
between all existing things to the
smallest scales with respect to all
existing things' standpoints in the
universe minus the universe that you
are making your perspective from and
that the earth that you stand on could
take the place of other things in other
universes, depending on the universe
that all things have their standpoints
at. other universes' time. This may
cause one universe's mass to
represent the space of another.

Perpetual Motion

In order for matter to be in motion, it
must have a partial state of being in
the form of energy and this form of
energy may transfer from one part of
the mass to another part and this may
be done by a principle known as the
uncertainty principle which states that
something in a constantly changing
position or velocity cannot be
determined as to the exact position of
it.

In order for anything to have an
energy state, it must have some form

of space and time for the energy to transfer, to move and to change in. This space and time could exist by mathematical means within the matter or other medium itself.

Once, space was thought to have a substance called an ether, which slowed all objects down eventually and didn't allow space to be completely free but this was disproved by Albert Einstein's General Theory of Relativity. However, one ether may exist and this ether is work and work can be the conversion of motion to heat or it can manifest itself in other ways, such as a motor converting electrical energy to mechanical energy. In such cases, the work done never completely converts the energy at any given position in space-time to the energy which some of it is being converted to and this is subtractive both of the energy that was converted and the remaining energy which was not converted and this produces some loss either way.

The result is that perpetual motion may only exist when the mechanism of perpetual motion moves in a free mathematical space that allows 100% energy conversion, which may prevent perpetual motion. On the other hand, if a device can be invented that can make use of all converted energies as well as transfer these converted energies in useful form to the other mechanisms of this device in a synergistic manner, then perhaps we may one day have a device with 100 percent efficiency and perhaps a perpetual motion device.

The Attraction And Repulsion Between Like and Opposite Charges

The field of a charged particle may
be considered the the particle itself in
a distributed form. The closer
something gets to the particle, the
more solid that it becomes.

There may even be an area around the charged
particle where the space is displaced
in such a way that the field may have
to make a teleportationlike
jump from the particle to a more distant point in
order to distribute.

When two like-charged
particles approach each other, this is akin to
the particles actually touching and
trying to move through each other,
causing a bit of compression.
The closer that they get to each other, the
denser that the field becomes and the
more that it resists, causing the
repulsion of the particle. The contact
between the fields of two oppositely
charged particles can be considered

akin to the two particles touching, so when the fields touch, then the only thing that doesn't cause them to touch in a teleportation-like manner is the distribution of the particles' position over space but this causes the oppositely charged particles to pull closer together and when this happens, the field, of course displaces more space and becomes denser which causes it to become more like the particle itself.

This causes the oppositely charged particles to pull together more forcefully until the particles actually touch.

Going And Coming Around

Space can be limited by folding it over time and observing it as only expanding or it can be stretched out from having this fourth dimension (There are possibly 11 dimensions of space according to Brian Greene, author of The Elegant Universe) which is time to being very large and possibly infinite. This may produce the boundary between infinity and being finite just as Stephen Hawking defined the universe as finite but with no boundary.

Also, there seems to be an equal and opposite reaction associated with space. every time something happens, it may send a wavelike signal through space by gravitational influence and by other forces. This 'signal' may influence the influenced body, particle or force to influence the source of the signal back just as much as it was influenced by the signal from the original source.

Black Holes, Relativity and Parallel Universes

If a black hole has any relation to another mass that isn't one, then that mass is the black hole, so the black hole first spoken of should end up scattering and surrounding that mass that is a black hole by relation.

This would indicate not only parallel universes but that for every body of matter, there is a black hole and for every black hole, there is so much of the exact same matter scattered around somewhere else. If these black holes or bodies of matter are from the other universe that that contains the reverse of what they are when they don't invert from relativity, then this should form a loop that if a parallel universe is entered by lets say me, then I would immediately come back to this universe. This loop would indicate how everything is relative by the inversion of their states with respect to the relative object.

Such a loop might be formed graphically with a computer programming language such as BASIC which is a very simple language and this might be demonstrated more easily this way. If anyone wants to program their computers to produce loops analogous to these ones if any of you can, do it!

If I am wrong, then it is out of the set boundaries of physics and mathematics by this universe, but if I am right then the computer should be able to carry this out.

The Cosmological Constant and Gravity

If matter produces gravity by warping space,
then shouldn't there be an initial wave
carrying a forward impulse associated with
it?

After the wave passes, then perhaps
between that matter and another object
there could be a reduced space? After that,
maybe a farther push and a farther reduced
space? If, so then this could be the answer
to gravitational attraction.
Second of all, it
seems like the more space there is that is
warped, then the farther out the space may
reach, but relativistically, this wave may
have the property of pushing something
through space instead of expanding it. At
the same time, maybe the pushing waves
as a result of more and more space
displacement get smaller and smaller and
maybe the stronger force of the reduced
space pulls one mass closer to another.
Between the pushing impulse and the
increasing reduction of space between, the nature of gravity as
being a wave may arise.

At the same time, maybe the reason for the
departing galaxies may be a great big push
from the initial space displacements and
after that, maybe the galaxies may rapidly
draw together from the gravitational
attraction.

The galaxies and other bodies of
matter may move farther apart in the same
way that driftwood is pushed to shore and
pulled back into the ocean. This may result
in a universe that accelerates in expansion,
but as the pushing waves become weaker
with distance, maybe this acceleration will
become less and less until the universe
stops accelerating in its expansion and
begins to slow down (decelerate) in its
expansion until it collapses.

Another possibility may be that the pushing nature of
the waves as the matter initially displaces
space may increase once again as some of
the space pushes into the matter and
produce some degree of equilibrium, which
may prevent collapse and also, after the
deceleration, cause the large bodies of
matter to move apart more and more rapidly
again.

This would produce a universe that
alternates between acceleration and
deceleration of its expansion.
On the other hand, the increase in energy concentration
from any collapse may help actually push
galaxies apart and the static relationship
between energy trying to push the universe
farther apart and the gravity trying to pull it
in may allow for the energy to push matter
farther apart.

It occurred in my thoughts the other day that if
gravity tried to collapse space, that the space
would store energy from compression that
would push masses farther apart with a force
equal to the gravitational force that they exert on
each other.

This would produce a universal

expansion in an attempt to cause collapse. This would allow for a cosmological constant to exist that would contain an equal force of anti-gravitational form to gravitational force and cause expansion of the universe at the same time and this would prove Albert Einstein correct in his theory as well as prove those wrong who made assumptions based purely on the expansion of the universe and no other factors. If I am correct, then Albert Einstein was correct that the anti-gravitational force is equal to the gravitational and wrong that the universe stays the same size. This theory would also explain why the large bodies of matter are getting farther apart.

Spatial differences, time and gravity

I was thinking about the idea of nothing ever being simultaneous, since the distance between all things requires for the occurrence to be propagated through space across that distance in order for any relation to occur. It occurred to me that this asymmetry or dissociation between points in space could be the reason behind time.

It also occurred to me that these differences could also indicate that the warping of space by all matter could be caused by these same differences by the oval circumference of the space depending on distance. These differences could be turned inside out to some degree by the overall mass of the object and turn a larger space into a smaller one and therefore bending it.

So, just as Albert Einstein said, gravity occurs when matter curves space.

Here is a good and simple experiment to demonstrate:
grab a long strip of paper and bend it to some degree and
then figure out what you have to do to turn some of this
bent paper inside out. What happens to the paper and
what does it look like?

If you can figure out how to turn
only a fraction of this bent paper the other way, you will
probably understand this theory more clearly .

You Are Where
You Are

This title is named after a saying in a bar by a friend named Pamela. If you are
familiar with Montrose, CO then you may know what I mean.

I was thinking about how many dimensions could form if
I treated the previous dimensions as flat and have found
that the mathematics behind this showed that there is a
limit to the number of possible dimensions when one of
the variables eventually turned the answers to negative
numbers when they were previously positive. As it turned
out, I found that there can be near 1,000 dimensions.
Keeping in mind that in order to fold a 2-dimensional
space into so many dimensions, the space would have to
become smaller and smaller and when reaching the limit
to the number of dimensions, I had a thought...It was that
if this number is exceeded, the number of times this
number is exceeded will cause a wrapping of all of
nearly 1,000 dimensions that would cause the formation
of as many more wrappings of such dimensions as there
are existing dimensions and this would square and then
cube and continuously exponentiate at a higher rate than
the previous dimensions could form a singularity and this
could cause some sort of spatial expansion which may
very well be the cause behind what causes the universe to
expand while at the same time, the folding of these

dimensions could be and form at the same time, the force
of gravitation and matter may be the frame that space
forms upon and since as previously mentioned in the
previous chapters of this book. This dimensional effect of
causing shrinking of all things spatial and expansion
might, if my thoughts are right, form the balancing act
between universal obliteration and the gravitational effect
of giving space its surface tension. If this whole theory
holds true, then maybe with each exponential increase
in the rate at which these dimensions wrap, the point at
which they do and and previous times at which they have,
there may even be exact duplicates of our universe and
exact duplicates of what is inside and since an equal time
has passed from the point that I this is being read back
the beginning of this whole dimensional occurrence, the
duplicates of our universe and the contents within should
all be of equal age as well as of all ages, producing a very
huge multi-verse.

This may cause anyone who attempts to
go back to the point of the wrapping that this particular
being is on, this being might end up going back to the
very beginning of everything, but because of the reversal
of the previously singular state, this being may end up at
the exact same point in space and time that this being
started out at during the attempt to do this. This is the
reason why I gave this chapter its particular name.
On the other hand, between these dimensional wrappings that
may occur new theory,accordingly with their tendencies
toward the state of becoming a singularity is a possible
indication that these wrappings may separate by staying only
withing the limits of present and past and the past may, in
theory, wrap up into the singular state just as a computer may
store information on a disk or sound may be recorded on
a tape and many physicists have considered the universe to be
a type of computer.

This may be a cause for the
directionality of time and another reason why nothing
can go into the past unless the state of being a singularity
occurs in a way that something is perfectly stationary
relative to all that surrounds, which is impossible if all
that is physical obeys the laws of physics.

Also, with each wrapping comes the greater length of
such wrapping and this wrapping may separate
completely from all other wrappings and causing the
universe other than a black hole, to stay only within the
present.

A black hole, on the other hand, occupies an area
of zero space and all space that surrounds it may just as
well be considered analogous to the two sides of a record
or other disc by the two dimensional area that it is small
enough to perfectly occupy.
The two sides could be considered two universes
and this may occur in all possible directions. This just might
also produce parallel, as well as identical universes.
Theoretically increased circumference of each wrapping may
also be a possibility of increased distance between all that
exists within the universe and thus, causing all large bodies to
increase in distance between, but this force may also have to
be on very large scales in order to work with enough power to
cause galaxies and other large bodies to speed farther and
farther from each other because of a possibly more
powerful gravitational force. As I have also mentioned
earlier, if the universe ever stops expanding, it would have
to concentrate all energy within into smaller areas and
this may or may not prevent the universe from collapsing,
but actually growing larger in the process while taking
energy away from all that is in the universe and eventually, all
masses all masses may eventually become supercooled.
The "Heat Death" of the universe simply states that the
expansion of the universe is what will cause all to cool
off until no energy is left to aid the expansion. This
theory on the other hand may be similar, but the cause
may be slightly different in its nature by the matter and
energy in the universe being consumed by the effort of
energy to push against the edge of the universe.
At this point, the universe store very large amounts of
this spent energy by converting it back to matter and
finally, in the end of the universe, this consumption of
energy may finally bring the universe back to the state of
being a singularity, which as previously mentioned, is
infinitely dense and has no size. With no space or time
around to allow for gravitational force to exist, this
singularity may unfold and once again, cause another
universe or multi-verse to form.

A New Model Of The Universe's Origin

I was busy putting up with the telepathic-like
'voices' of my paranoid or undifferentiated schizophrenia, the
diagnosis being unclear to me, since one hospitalization ended
in a diagnosis of paranoid andthe second hospitalization 7
years later ending in a
diagnosis of undifferentiated schizophrenia, last night
when one of them ended in an interesting
insight as I discovered that there seems to be a pattern
consistent with my scientific writing and that is that
when I write down my insights, my voices fade and I feel
more clear in thought. The insight was very
interesting and may very well be a plausible model of the
formation of the universe that also allows the two
theories relativity, which are called General Relativity
and Special Relativity, the latter mention in order of
which was first and secondly thought up by Albert
Einstein, to still be correct as well as for quantum physics
which had its beginnings as Albert Einstein's proven
theory that electromagnetic radiation in all
frequencies of oscillation could exist as a wave and as a
particle. This is known as the wave particle duality.
As my theory begins, what could happen to a void
having the nature of the absence of all existing forces and
things if even the most infinitesimal of masses displaced
an area of such absolute absence?
My theory is that if this could happen, then the void would
itself become a force by itself being present around this
infinitesimal matter and literally warp around it in a manner
which i could form into something existent, but also allows
for the motion of all presences from inside moving
around with no resistance to motion what so ever and in
this instance, this could be space. So, with the
eccentricity of the former void curving around this tiny
mass, how large could this void be? My mathematics

indicate that the overall quantity of what formerly existed as a void being divided by pi and then divided by the smallest infinitesimal value possible without becoming zero, this would be a number that would almost be infinite but not quite, but would be very large and indefinite. The reason by which I only divide this huge former void by pi and not to its third power is because it is already multidimensional with the originally given 4 dimensions, now considered to be 11 dimensions, so pi needs no exponentiation. This curvature may also possibly in the process of compression, squeeze the spent energy out like a rag being squeezed and its water being removed and this spent energy may manifest itself as matter or could have.

As the impulse of existent matter emerging in this universe, the very presence of it could curve more void with greater force and exponentially increase the accelerate expansion of this universe and cause the universe to form matter more rapidly with exponentiation multiplied by acceleration, producing more and more matter at a time but with the occurrence of increased space and time, the rate at which matter forms should, by mathematical principle, be a constant when applied to matter formation for every cubic unit of space as well as time. This theory allows gravitational force to be still considered a result of space curvature by increased warping of space in one direction from one piece of matter and to the next and the same in the exact opposite direction from the latter, showing some kind of resistance to its own curvature which would cause an impulse that would push the pieces of matter together. Secondly, the increased circumference of space when placed a rather long distance from the given point in space and time should allow for a good explanation for which it is observable that the larger, more visible bodies of matter, such as galaxies are moving farther apart. This is assuming that my new idea indicates that time is a replacement of one segment of space by new ones as space grows larger and this may allow for exact recordings of what has happened in the past by propagation of change in position or state in a way that resembles a wave and an explanation of why electromagnetic waves propagate through space when an

electric or magnetic field is placed in
back and forth motion.

A good classic example of this
would be the polarity reversal of alternating current
producing electromagnetic waves which is the cause of
radio interference caused by nearby power lines
and if a large set of power lines is close by, especially if it
is a substation or a power plant, the electromagnetic
waves can be more if not highly concentrated and cause
almost an impossibility for good clarity of the
reception of a radio or television station.

This theory also eliminates the need for there to be any
large and let alone infinitely dense masses known as
singularities as found in black holes to be present for a
rapidly expanding universe to occur especially
when this theory entails what resembles a falling row of
dominoes that fall faster and faster as more dominoes fall.
If anything is moving in free space, the pieces of space
being taken the place of by the next also become closer
together and the time periods shorter in such a setup,
which coincides with Albert Einstein's theory that time
can stop at the speed of light. Time will stop if something
moves fast enough that it touches every new segment of
space as soon as it forms just like the sound
waves become highly compressed if an airplane or jet
moves at the speed of sound, but no faster. If the
airplanemoves no faster, then the sound waves as they leave
should return to their original size and pressure as they
leave, provided that the speed of sound is not
exceeded and the jet or airplane should sound like the
typical airliner from the ground, but if this speed is
exceeded, then the airplane will produce the so-called
sonic boom, which results when the sound waves
remain compressed as the leave when they are penetrated.
This could be caused by the fact that everything is
relative and if they are moving in one direction at the
speed of sound, the waves have a lesser chance
of decompression from the state of being a shock wave,
because the speed at which they can widen in the
opposite direction is reduced and the pressure still may
remain high, and the shock wave should remain a

shock wave.
If the speed of light was exceeded, what do
you think would happen if the light passed over
every emerging segment of space at once?

Personally, I would say that the universe would be in a collapsed state
of infinite density and appear to be a singularity if I
moved past the speed of light. That is what may
be required to exceed the speed of light. Anyway, this is
my new model of the formation of the universe.

Universal Quintessence and other Universal Occurrences

Since the universe started out as an infinitely dense
particle called a singularity, then without space
time for gravity to exist or possibly similar and
integrating forces to hold it together, then it seems to me
that this singularity must have exploded with infinite
force or close to it.

There would probably be a
dividing factor that changed with one number to stay as a
constant for the rate that the universe
accelerates. The singularity, I think may have been
simultaneously in a state of kinetic energy and
perfect rest.

This may probably allow spent energy to
become matter in this universe while the infinite
initial force at the beginning of the expansion is divided
by the mentioned constant number which
would indicate that the universe should expand at the
value of the infinite force divided by the overall
expanse of this universe, as mentioned before and this
answer may indicate the overall potential for

energy in the mass of the overall universe as well as the speed of light. This would probably mean that the famous equation by Albert Einstein "E=MC^2" may be equal to the mass multiplied by a distributive number of space over the distributive number of time when space and time are equal arguments and when time is zero and space is any significant figure, in this case, the speed of light. So, when time is no longer distributive and becomes zero, then motion through space over zero time would actually mean mass multiplied by infinity. When space is no longer a distributive function, then matter should become rest mass that is perfectly stationary, but then it would probably become a wave trying to decide whether to occupy time or space and move at the speed of light instead and the same applies for time. So, without one or the other, the mass would probably move at the speed of light.

As black holes form and therefore open more room in this universe for matter, I think that maybe the new matter forms at the same rate as a black hole crunches matter into singularities. This, based on older Blogsthat I have placed may indicate that because of smaller space or at least space with greater curvature ofspace-time may cause the universe to attempt a collapse, but the inner energy might push the universe outward, due to concentration of energy. Because of this, it, to me at least, the farther expansion of the universe may cause more of the unspent energy to form matter more rapidly and this matter may produce more black holes and as the black holes multiply, the universe may try to collapse, but because of more black holes present in this universe, more energy may be spent and turned into matter and as a result, more space-time may be produced and this may explain why galaxies are moving farther apart.

Black holes may serve to cause the universe to accelerate in expansion by the occurrence of greater matter being produced in this universe and as the energy is spent, the universe may expand faster than a previous time.

As a collapse is attempted by this

universe, then the energy may be more concentrated than the previous attempt to collapse. So, between attempted collapses and the occurrence of spent energy, the attempt to collapse may serve to keep the universe from ripping everything apart that allows it to function and the spending of the energy may aid the expansion of this universe.

Between multiplying black holes and the production of matter, and spending of energy, these may work mutually to accelerate the expansion of the universe. However, accelerating rapidity of universal expansion might cause the energy to be spent more rapidly and in this case, energy just might run out and the universe might eventually collapse back to a singular state as all matter draws together and as this happens, the matter may start forming black holes and then more massive black holes as they collide and finally, the universe may actually collapse into the original singularity. Finally, galaxies and other larger bodies of matter may be the most central points where the new matter collects, since a black hole may in a similar fashion to galaxies in gravitational nature and universal gravitation may be the reason behind the rapid transfer of matter from the edge of the universe to places where matter collects; These collectors of matter can be just about any large mass...galaxies, solar systems, black holes, large rocks,.

Energy, Matter, Waves, Particles, Space and Time

Space is the medium of which energy travels through. Time is an inverse value of space that is regulatory of the concentration of energy over this space and time is also the inverse value of space.

Because time is the inverse value of space, the actual wavelength of the emitted energy between the beginning and end or the crest and trough of the wave could be determined by the time taken to complete propagation of a wave over a distance equal to the wavelength itself. As you may have insight on this, the distance between crest and trough are literally half of the wavelength itself and paradoxically the amount of time required for completion of the oscillation of the wave.

The amount of time required for energy to be released may therefore, be the potential for a specific amount of work done over any given period of time within the limits of the length of the wave.

The amount of any work done at all cannot occur without the presence of matter and its conversion to energy. This in simple terms indicates that a burning candle under normal conditions will do so very slowly over a long period of time, but if this candle burned all of its fuel within a shorter amount of time3, then the4 flame would be brighter as well as hotter, because of the more rapid release of energy in this shorter time period. If an average birthday candle could use up its fuel within a millionth of a second, then it would most likely explode and blow up the surroundings with great heat within a very large radius.

This would also indicate the huge difference between radioactive decay of a piece of uranium or other radioactive substance and the sudden splitting of the nuclei along with much more rapid equivalent to decay that by comparison of the 4-billion year half life of uranium and the sudden release of the energy in an atomic bomb that could nearly equal all of the energy released during the half-life of uranium, which may occur within milliseconds or less in this case.

I also have a similar idea that involved the explosive force that may come from the exposure of spontaneously combustible fluids to large and pressurized quantities of oxygen in a small space. If enough oxygen could be pressed against linseed oil to burn a liter of linseed oil with a 2-second time period, this would be a huge release of energy that would constitute an explosion.

At the same time, this scenario could be a very good analogy expressing short wavelengths as having higher energies in a smaller space because of their potential to cause change or do work in a shorter period of time.
As a concentrate, these waves are a particle with accordance to Albert Einstein's proven existence of waves simultaneously existing as particles.

This proven theory also expresses the possibility that fields of any force are particle in concentrated form but are waves in the nature of a field. combined within and over a specific volume The existence of fields in particle form is called the quantum field. The quantum field viewed from perspective of uncertain position may behave as a wave, but by potential exertion of force, may be viewed as a particle.
This indicates to me that light and radio waves along with the rest of the electromagnetic spectrum have properties that are identical to field properties with the exception that they only reach a limited distance.

Gravitational force can act at the same distances as light and radio waves, but obtain their wave properties by the inverse relationship between space and time and with each inversion, there is one pulse generated in the form of gravitational particles known as gravitons, but the inversion back the other produces the graviton particle and gravitational wave as well. This may produce a folding of space responsible for Einstein's inclusion into relativity that gravity is a result of matter curving space. The greater the slope of the inversion, the less occupation of space and time both and the greater the gravitational time dilation that occurs. I am not sure who proposed time dilation first, but I see a coincidence between gravitational time dilation and inversion of space and time that causes space and time to fold in such a way that curvature by matter makes space and time one and the same.

-Nate Durham May 27 2008 10:30PM

Gamma Ray Bursts and Black Holes

I had a hunch the other day that based on an earlier theory of mine that the gamma ray bursts observed in outer space lately may actually be a result of the high concentration of energy by the gravitational collapse of a star during the formation of a black hole. The high concentration of energy associated with gravitational collapse may by the effect of space and time being less displaced at a distance, cause the frequencies of electromagnetic emissions such as Hawking radiation to become lower and for the waves associated with these frequencies to become longer as there may be an allowance for events to occur farther apart due to higher allowance for distance between these points.

The frequency shift (known as a red shift when frequencies become lower) may occur with direct proportionality to the distance from the center of the gravitational collapse as well as change in the strength of the gravitational field. This may provide a partial explanation for

why it is possible that space and time could be
highly connected.

Also, increases in frequencies are known as
blue shifts. I thought that I might as well throw that in.

The Universe and Formation of Time

If the universe started out as a singularity, which is an
infinitely dense and zero-size, but significant mass then it
seems to me that the distribution of the occurrence of the
explosion of the singularity over the space that it produced
and continues to produce would by this distribution define the
nature of time.

If this is true, then as the pieces of the singularity scattered
and still scatter would have a kinetic energy as well as a
temperature that would be inversely coincidental to the
distance fro the point at which the singularity once was which
this point would behave in a way very similar to the way that
a bubble formed by soapy water would have behaved.

This bubble like behavior would consist of the difference
between surface tension required for the bubble to take form
as well as the elasticity produced by broken surface tension
which would be required for the bubble to inflate.

Gravitational time dilatation may be a result of the differences
in time requirednto get from one place to the next in the
presence of no mass divided by the mass of pieces of matter
involved in the gravitational interaction between them.

Gravitational force may be the result of formation of surface
tension of space around a mass and the breaking of this
surface tension by another mass. This would reduce the
distance between two bodies of matter.

As a result of this, time would also be less. At the same time,
however I believe that time by origin is a result of the distance

between the occurrence of the explosion of the singularity and the distribution of such an occurrence over the space that it formed itself.

Gravitational differences, because of time dilation may increase temperatures of masses in mutual presence. This may possibly indicate that some electromagnetic radiation may radiate from areas where there is change in gravitational time dilation, such as by the acceleration of falling matter. Electromagnetic radiation is another word for radio waves, microwaves, infrared, visible light, ultraviolet, X-rays and gamma rays and these are placed in sequential order from lowest energy and longest wavelengths to highest energy and shortest wavelength.

The time differences in all particles and bodies of matter might be compared to the rainbows that are seen moving around in bubbles of soap and water.

CERN, Black Holes and Electromagnetism

After watching a show about the very large collider at CERN, I begin to wonder if black holes other than what might have been at CERN might produce some degree of electromagnetism in a unified form. If so, then because of what is known as super-symmetry, this may very well occur, since singularities may be capable of looping space and time.

I have a hunch that when the collider was running, the singularity may have literally stretched a singularity in a circleand broken it down into its constituent energy and thus, forces.

The gravity may have been accelerated until space could loop a magnetic field. If all of this is possible, then black holes may be very short lived, since singularities would have the same effect and the energetic result possibly may be that the singularity and super-strong electromagnetic field may be one and the same thing.

The result may be that the black holes may explode into pure energy at the instant that infinite density is achieved. This could kill a lot of life, but luckily, we are nowhere near a black hole.

The gravitational time dilation during the formation of a black hole may be responsible for Hawking Radiation by increasing the motion of atoms and other matter as well as increasing the energy of the matter by bending both time and space so that both velocities can be mutually multiplied. The result may be the increase of temperature in a manner different from the accepted theory that friction between atoms does this.

Multiplying matter by the velocity during time dilation would therefore be the same expression as E=MC^2. Albert Einstein was of, course the dreamer of time dilation and E=MC^2, but in no textbook could I ever find a relationship between temperature and time dilation. The frequency shift (known as a red shift when frequencies become lower) may occur with direct proportionality to the distance from the center of the gravitational collapse as well as change in the strength of the gravitational field.

This may provide a partial explanation for why it is possible that space and time could be highly connected. Also, increases in frequencies are known as blue shifts. I thought that I might as well throw that in.

I had a hunch the other day that based on an earlier theory of mine that the gamma ray bursts observed in outer space lately may actually be a result of the high concentration

of energy by the gravitational collapse of a
star during the formation of a black hole.
The high concentration of energy associated
with gravitational collapse may by the effect
of space and time being less displaced at a
distance, cause the frequencies of
electromagnetic emissions such as Hawking
radiation to become lower and for the waves
associated with these frequencies to become
longer as there may be an allowance for events
to occur farther apart due to higher allowance
for distance between these points.

The Universe and Formation of Time

If the universe started out as a singularity, which is an
infinitely dense and zero-size, but significant mass then it
seems to me that the distribution of the occurrence of the
explosion of the singularity over the space that it producedand
continues to produce would by this distribution define the
nature of time.

If this is true, then as the pieces of the singularity scattered
and still scatter would have a kinetic energy as well as a
temperature that would be inversely coincidental to the
distance fro the point at which the singularity once was,which
this point would behave in a way very similar to the way that
a bubble formed by soapy water would have behaved.
This bubble like behavior would consist of the difference
between surface tension required for the bubble to take form
as well as the elasticity produced by broken surface tension

which would be required for the bubble to inflate. Gravitational time dilatation may be a result of the differences in time required to get from one place to the next in the presence of no mass divided by the mass of pieces of matter involved in the gravitational interaction between them.

Gravitational force may be the result of formation of surface tension of space around a mass and the breaking of this surface tension by another mass. This would reduce the distance between two bodies of matter. As a result of this, time would also be less. At the same time, however I believe

that time by origin is a result of the distance between the occurrence of theexplosion of the singularity and the distribution of such an occurrence over the space that it formed itself.

Gravitational differences, because of time dilation may increase temperatures of masses in mutual presence. This may possibly indicate that some electromagnetic radiation may radiate from areas where there is change in gravitational time dilation, such as by the acceleration of falling matter. Electromagnetic radiation is another word for radio waves, microwaves, infrared, visible light, ultraviolet, X-rays and gamma rays and these are placed in sequential order from lowest energy and longest wavelengths to highest energy and shortest wavelength.The time differences in all particles and bodies of matter might be compared to the rainbows that are seen moving around in bubbles of soap and water.

Energy, Matter, Waves,

Particles, Space and Time

Space is the medium of which energy travels through. Time is an inverse value of space that is regulatory of the concentration of energy over this space and time is also the inverse value of space.

Because time is the inverse value of space, the actual wavelength of the emitted energy between the beginning and end or the crest and trough of the wave could be determined by the time taken to complete propagation of a wave over a distance equal to the wavelength itself.

As you may have insight on this, the distance between crest and trough are literally half of the wavelength itself and paradoxically the amount of time required for completion of the oscillation of the wave.

The amount of time required for energy to be released may therefore, be the potential for a specific amount of work done over any given period of time within the limits of the length of the wave.

The amount of any work done at all cannot occur without the presence of matter and its conversion to energy. This in simple terms indicates that a burning candle under normal conditions will do so very slowly over a long period of time, but if this candle burned all of its fuel within a shorter amount of time3, then the flame would be brighter as well as hotter, because of the more rapid release of energy in this shorter time period. If an average birthday candle could use up its fuel within a millionth of a second, then it would most likely explode and blow up the surroundings with great heat within a very large radius.

This would also indicate the huge difference between radioactive decay of a piece of uranium or other radioactive substance and the sudden splitting of the nuclei along with much more rapid equivalent to decay that by comparison of the 4-billion year half life of uranium and the sudden release of the energy in an atomic bomb that could nearly equal all of the energy released during the half-life of uranium, which may occur within milliseconds or less in this case.

I also have a similar idea that involved the explosive force that may come from the exposure of spontaneously combustible fluids to large and pressurized quantities of oxygen in a small space. If enough oxygen could be pressed against linseed oil to burn a liter of linseed oil with a 2-second time period, this would be a huge release of energy that would constitute an explosion.

At the same time, this scenario could be a very good analogy expressing short wavelengths as having higher energies in a smaller space because of their potential to cause change or do work in a shorter period of time.

As a concentrate, these waves are a particle with accordance to Albert Einstein's proven existence of waves simultaneously existing as particles.

This proven theory also expresses the possibility that fields of any force are particle in concentrated form but are waves when combined within and over a specific volume and/or area.

The existence of fields in particle form is called the quantum field. The quantum field viewed from perspective of uncertain position may behave as a wave, but by potential exertion of force, may be viewed as a particle.

This indicates to me that light and radio waves along with the rest of the electromagnetic spectrum have properties that are identical to field properties with the exception that they only reach a limited distance. Gravitational force can act at the same distances as light and radio waves, but obtain their wave properties by the inverse relationship between space and time and with each inversion, there is one pulse generated in the form of gravitational particles known as gravitons, but the inversion back the other produces the graviton particle and gravitational wave as well. This may produce a folding of space responsible for Einstein's inclusion into relativity that gravity is a result of matter curving space. The greater the slope of the inversion, the less occupation of space and time both and the greater the gravitational time dilation that occurs. I am not sure who proposed time dilation first, but I see a coincidence between gravitational time dilation and inversion of space and time that causes space and time to fold in such a way that curvature by matter makes space and time one and the same.
-Nate Durham May 27 2008 10:30PM

Gravitational Interaction and Time Dilation and the End results

I was thinking and I think the warping of space and

time may more literally be space and time actually folding and this may define gravitational time dilation better than any book I have read, that is if it is true. Because this theory allows time to be the equivalent of Space.

It can very well define gravitational force as being a force that also allows energy to be its own equivalent in matter and vice versa. This would also explain Hawking Radiation in the sense that differences in time dilation can alter the speed at which something is moving or in other words, concentrate the potential for matter to become energy and the closer it gets to high gravity bodies of matter, the higher the kinetic energy and and the hotter it gets. This might cause matter to become pure energy or in other words, electromagnetic radiation at the event horizon of a black hole by the lack of time with reference to surroundings. Thats my only news for now.

Parallel Gravitational Interaction

I was also thinking that maybe gravitational fields may possibly never be identical, because then space and time would become parallel, but then again, this may be what drives things farther apart in this universe by gravitational interactions of identical strength and therefore the repulsion by the fact of being parallel in space and time

Motion Might Bend Space as well as Matter Itself.

Nate Durham March, 14, 2008

I know that they say motion is relative and that matter warps space. I also know that the distance that something travels over time is the speed or velocity at which it moves. What if the dimensions of time and space could become one entity by the fact that the distance over the time would equal the distance traveled?

If this is true, then motion must bend space and time so that they overlap and are one. If this is true, then it would easily explain WHY things are relative.

Gravity warping space is just about the same as motion bending space and the distance away from the source of gravitational force should therefore define a lessening of the bending of space by the fact that the greater distance increases the time that it would take to get from point A to point B. This would lessen the acceleration of an object toward the gravitational source as well as explain why things slow down against gravity by the fact of space and time unfolding and therefore stretching the amount of time that it goes through as it moves farther away. Its the same mass as before, but less of it would be energy at this point.

Electric fields could by relation to a passing object, also be bent identically to the amount of space that is displaced, but also the strength of the field might multiply by the curvature of the field to produce not only a magnetic field, but a tension of space and a unified electromagnetic field by the 90 degree angle that electric and magnetic fields sit from each other in relative position.
The motion of an electric field should also bend space and thus, bend the electric field and produce an electromagnetic one.

Remember the electromagnets in elementary science classes?
Well, I think this could provide a good explanation for that
particular phenomenon. The energy in motion may define the bending of space as a
measure of matter that might be responsible for the bending of
space as well.
Nate Durham March, 14, 2009

Second Thoughts about Black Holes

When matter is in fact in relative motion, then by the time a singularity can form, I think matter would already have reached a state of infinite energy by the fact of infinite time dilation. In this case, though there is evidence of an existing singularity, this may not be true if a large mass of high density only exists before the moment of infinite time dilation. So, a singularity may possibly prove to be impossible unless infinite energy is the trigger. Evidence of the existence of a singularity may be wrong, since the super-gravity can only be taken into consideration when matter has not reached a super-energetic state in which time dilation multiplied by the mass would probably equal the energy that would fly away from the center of the black hole that a singularity forms.

Depending on the speed that the mass flew in with and the degree of time dilation involved, this may determine the point outside of the center of the black hole singularity that the mass reaches an infinite state of energy which gravitational time dilation would therefore produce an infinite red shift, thus lowering the frequency of and increasing the wavelength of the high energy particles. This may mean that though evidence of a black hole may exist, infinite gravity might not and this may provide an explanation of why the laws of physics break down at the point of infinite gravitational force.
The inner mass of a possible singularity would most likely be passed on in this case as a burst of energy that would pull the high energy inward, accelerating the formation of energy farther as on big explosion. Could this be an explanation for the gamma ray bursts? If so, we do await our fate, but it may be a long way off, hopefully.
Pure energy has yet to prove what actually caused the formation of the universe and the only way I can see plausible for finding it out would be to dissect a photon and break the physics of what is called super-symmetry which at this point would take an infinite equation beyond what can be computed by a device which must abide by physical calculations.

How Far Can Calculation Take Us?

Energy is what is a looked for means of detection by its emission in the LHC. Dissecting photons would be an interesting subject because to dissect pure energy would be to produce a super-fluid with more dimensions to dissect the more it is dissected, giving the universe an infinite complexity in the true makeup of all things and at the point that it hits an infinite elasticity due to progressive complexity with breakdown, this may be where super-symmetry sits and in this case, physics and mathematics become one strand in helix form. This would mean un-breakability of supersymmetry and this leads me to the next question. Does DNA contain dimensions that we don't see and are the differences in structural makeup also a variable in the dimensional makeup and in this case, is our conscious makeup and individuality determined by the differences in dimensional structure in a way that the possibilities could be nearly infinite and in this case, was life a predetermined to occurr on the basis of the mathematic and physical paradox of super-symmetry between what is physical and what is mathematical? If it was, then it is definite that there is an infinite possibility to who we are and what we are in our consciousness! I am proud to be a mathematically and physically predetermined individual in a way that nobody but myself can ever know me from the inside out! Hey, everyone

don't believe everything you see in group mentality, because your true makeup is predetermined by what governs the universe and you have your place in life and not even the most powerful computer will never compute this!!!!!! You are predetermined as a self and not a somebody else.
Your personality and identity are a shape and they are the frameworks from which you grow on while as a vine of morning glory, your individual inner self is the chain link fence by which you define yourself with by your own personal growth.

From matter to energy, true progression is to keep learning and be on a quest for knowledge while we all know absolutely nothing but what we learn from the outside to the inside and not from the inside to the outside. True knowledge is from beyond and learning is from within. I like physics because I like to keep learning and grow and the quest for knowledge is never ending in the belying nature of all that belies while a politician can only know so much before true complexity ends his or her quest for truth. As Albert Einstein said, "Politics are for the present and equations are for eternity."

I know there is one out there and I know it is beyond the realm of physics and I know that the most powerful politician will never be this or have grasp of this concept but to hold this exact phrase by documentation, never to be able to manifest him or her self as being such a concept as a physical being. As beings, we are spiritual and only physically are we ever matter.

Dielectric heating VS. The Opposite Effect

Dielectric heating may turn another page for me and it seems to reveal that Dielectric effect can also allow the atoms of an electrically non-conductive gas to invert in position upon the reversal of the polarity of a physically present electromagnetic field, compressing the gas and causing the atoms or molecules of this gas to give up their thermal kinetic energy.

As this occurs, there might be an opposition to compression of the gas by its thermal kinetic energy and in turn, giving up its thermal kinetic energy by compression may allow farther electromagnetic polarity reversal to compress the gaseous substance even farther and cause the atoms to give up even more energy. Finally, the gaseous substance involved may liquify or even freeze from the energy that is lost by the atoms and/or molecules that compose that particular gas.
The inversion of the position of the electrons of an atom may also produce a more localized area of the same occurrence within itself, but only in the presence of an electromagnetic field or alternating form. This could easily cause these atoms to also absorb the energy within reaching distance of the electromagnetic field in quantities that are the same as the energy involved in the loss of temperature of these atoms and this may indicate that actual work is used to super-cool the atoms that make this gaseous substance up. Therefore, adding these inverse energies as a sum would be equal to zero and indicate the input truly equals output in any energetic occurrence. This, of course is a good indicator that energy is always conserved in one way or the other and none is ever lost in this universe, but it goes somewhere else instead.
This may be responsible for the Bose-Einstein Condensates to form around high-speed microprocessors, which constantly produce rapidly changing electromagnetic fields as a byproduct of the function of a microprocessor in a computer, whether it is your laptop or a supercomputer in a huge laboratory. A Bose-Einstein Condensate is a very cold gas that based on Albert Einstein's theory and not relativity, but a different theory, collects at an equal quantum level when it is extremely cold.

Matter, Energy and The Concept of

Space and Time

Every 6 years is a perfect time period. 12 years, foundations are shaken. Every 24 years,there is

For every same time passing, it happens again like Thanksgiving!
This is the mathematical biorythm for a universal time line.
Make it seconds, minutes, hours, days, years, decades or centuries! Look at it and think! Go back X years, X months, X days, X hours, X minutes or X seconds or milliseconds even or all of the above together and you will have documented past instantly!

Based ont he fact that everything is relative, then under any gravitational condition of spatial curvature this will not change. These numbers are a constant.

Theoretically speaking, the magnitude of an event is based on its resonance with other events in time and this could cause what is relative to increase on the quantum mechanical scale. The law of attraction is a resonant principle in other words. This is much like karma. Everything builds up in its magnitude the next time around. Each time, it amplifies in significance. The curvature of space may not just gravitate objects, but gravity may have the effect, just by curving space, gravitating occurrences in time and grvaitation data like an efficient supercomputer. The singularity in a black hole in other words consists purely of quantum mechanics.

This is the same as we exist in our own realities. We walk in our own minds.
My conclusion here is that quantum mechanics and relativity are relative in angle.
In unity, as I posted at the International High IQ Society discussion boards,
"Matter is a processor, time is a process, space is the processed and energy is the action." On the other hand, time may not exist with the inspiration of a friend named Cindy Zaccagnini.
In fact, as a purely energetic substance, space may be the energy that composes light with one difference. The space may have to be disturbed and flung in one direction in order for this to happen.

Quantum fields may be proven this way. One example is gravity. The displacement of space by gravity may cause the spatial potential energy to push away and then to neutralize as an energetic substance wit the exception that the energy of 'greater space may rush in from the pressure differences in order to fill this void in and literally suck bodies and particles of matter and energy together depending on the magnitude of displacement and action of force on the space in the first place. In a black hole, for example, the energetic element of space rushing in may be what makes it difficult for light to escape as a result of the immense violent displacement and disturbance that gives a black hole its ability to withhold light.

The formation of photons may be no more than the directional disturbance of the energy that makes up space in the first place.

In this case, the inverse square law may be what determines the difference in amplitude and frequency in such things as lasers.

The fact that frequency or energy increases when a source of light is moved in the same direction as its emission may be a result of decreased scatter in disturbance with decreased distribution of the disturbance of the energetic element that makes up space.

This would also explain why faster moving things age less quickly relative to surrounding
similar objects moving at lower velocities and so forth.
The accumulation of matter on an initial mass at a given increase in velocity may very well be the accumulated energetic substance that makes up space upon this particular mass.

The formation of waves may be no more than the action of disturbance of the energetic space in scatter formation with accordance to the Inverse Square Law. Albert Einstein in this perspective was right in his measurements and wrong in the concepts of space and time. However, similar things become resonant and increase in amplitude each time they happen so be warned!

A Chapter of Random Thoughts Otherwise

Electromagnetic Cannon
by Nathaniel Durham
11/04/2009

I have been thinking for several days. If I could project a magnetic monopole a fair distance, then since it would be mono-polar, then it could lens reality by the bias that reality contains and we could see the insides of what makes our reality up by the possibility that it this could turn the localized area of electromagnetic occupation inside out. As a result, we could probably see not only what makes up our deepest thoughts, but what literally makes up the insides of what makes them up to our maximum mental capacity, which theoretically is unlimited. Could this be a cure for mental retardation and a social door for autism?

Such a device would involves one magnet and one electromagnet and one long set of electrical pulses to provide the 90 degree rotation or the dip that would project hyperbolically at the other end of the electromagnet. A guiding device, such as an aluminum pipe or a pipe of other metallic or metallic-like and non magnetic, but electrically conductive material would be just fine as a barrel for projecting the 'Electromagnetic Smoke Ring' this since the skin effect of the metal would be a great cushion for the travel of this mono-polar lens that would resemble a smoke ring with similarity. Electromagnetic induction would most likely throw lightning-like discharges from the metallic barrel. On the other hand, insulating materials would be enhancing of the lensing effect on reality. If a solenoid or a properly arranged set of electromagnets was placed along the barrel, the lightning-like discharges may possibly be focused like the original laser and be directed at a target.

Finished around 4:50 PM
Modified at 10:39 AM on 11/06/2009

A Ring Around A Black Hole
By Nate Durham
11/04/2009 9:11PM

The particles that produce fields should contain the same mass as the field. Everything is relative, so this is massive dispersal of energy over larger areas of space and this makes things relative and if the spatial displacement is energy, energy is matter. The laws of attraction are based on density, so electrical attraction should have increased field density and repulsion should have decreased field density, but that is of, course on facing sides. The sides facing away should be the opposite. Strong nuclear force is almost pure solidity and should nearly equal the mass of the particle which would explain for example, why splitting atoms releases so much energy. Gravity on the other hand should disperse more rapidly and have less energy concentration with distance. Uncertainty is probably determined by the rapidity of dispersal of a given force.

As the universe expands, the particles disperse more rapidly and therefore giving rise to an explanation of why everything is getting farther apart. On the other hand, the resistance to kinetic change, inertia, that is, probably gives a driving force to the expansion of the universe as space pushes against matter and drives it forward at the same time that a resonant effect pulls matter together. Super-massive black holes will likely cause the stars in spiral galaxies to form tight rings around an apparently unoccupied area of space where attraction meats its opposite. There should eventually be galaxies that appear to have ripple formations consisting of stars and other bodies.

I wouldn't be surprised if the Hubble telescope eventually spots this.
-Some time probably at night, last week.-
Reality is a reflection of our mind. It is like a fluid because we can disturb it, but if i had pure solidity, it would not change and we would not think or feel anything but one thing and our minds would be frozen. Reality also has the tendency to have different levels and assumes a fluid-like nature because it has the assumption of different levels that make up our consciousness and therefore, it's exact basis is energy and our consciousness is therefore driven by energy which is why we must always feed our minds.

If we did not stir up our realities, our realities would eventually solidify and we would learn nothing. We learn through all of our senses. Everything we sense will eventually cause us to think if we are not in a state of constant thought already and just don't know it.
This is probably why the same thing over and over again can drive us mad. We as learners don't like to solidify in our realities and that includes thinking.
Boiling this down, we would not truly be living if we did not think or in other words, stir up our realities, so we must be creative to be conscious and live.
Also, not only do we need creativity to stir reality up, but we need to be able to stir reality up in order to be creative. If we only had intelligence to rely on, then we would not be able to be creative and we would only think in pure concretion and our realities would solidify. I believe that we all share a common consciousness and that this consciousness can only be capable of stirring reality by viewing from different perspectives.

From each perspective of every being, I believe that we can form each others' reality and stir each others' reality by intellectual intercourse of ideas to form new ones about our inner selves as well as our surroundings. This is probably why it is easier for similar intellects to socialize mutually than to talk with a dissimilar intellect, but even then, the lesser intellects need to feed themselves with some information and ideas from the greater of intellects.
Edited around November, 20, 2009 7:55 PM

To make good, we swing to the opposite side of our evils. In times of oppression, we beat our assailants. In times of assault, we become assassins.

We are composed of 2 evils. One see reality by cross reference and one is delusional. We swing use one evil to defeat the other by turning evil on itself, which composes the thinker. The creative thinker doesn't need a mirror to know that he or she exists, but the delusional thinker always needs a mirror to see that he or she exists at all.
The paper and the pen are the best reflection of thought because they can have the reflective capacity to confuse a quantum computer with paradoxical nature. In these times, we could all use the paper and the pen. This gives the balance between good and evil.
Edited on November, 20, 2009 7:00

Overthrows are a great subject and when I overthrew my oppression, I had intellectual freedom and when I had intellectual freedom, my oppressor was overthrown and I mastered my mathematics and thoughts.....It was a great goodbye to CJHS and its corrupt ways under the RE-1J policy of hating dead poets.
From my Facebook.com profile. Right arouind 8:00 PM November 16, 2009

Every move on a chessboard is every other move from every other perspective. You can never be another individual, because if you were the black piece from your perspective, you would be the white piece from the other. Don't assume you are another individual.
November, 16, 2009 8:09 PM

The future is a complex analogy to the present and the present is a complex analogy to the past. The past is singular and the future is inflationary in its model but it is symmetrical to the present by structure. The present wraps around the past and the past is a singularity while the future is a point.
November, 19, 2009 5:31 PM

There was an amazing fat burning pill that could cause someone to burn 300 pounds of fat in 10 minutes. That poor soul forcefully urinated and vaporized instantly. This is a big lie, but it is a fact.
November, 16, 2009 8:28 PM

A wall of stupidities is vulnerable to an explosive intellect and when the wall breaks, out flows the tsunami. One blast and it knocks them speechless.
November, 16, 2009 9:24 PM

If an alternating current has a choice to either flow across one gap or the other gap and switches direction, then logically, it will follow the alternate path and if it follows the alternate path, then most likely, it will jump between alternates by induced potential and form a triangle of electric arcs. I am in my alternate right now and I am thinking. Now, Igor......See what happens when the flourescent is placed between the

alternates!
November, 16, 2009 10:30 PM

I don't capitalize on my books, I socially distribute them. Thats why I give my friends free copies of my files! It's a friendly gesture and only those who I trust with my ideas will get them for free! If you are trusting with your ideas and give them out to just anyone, you just might get screwed! Keep your ideas in good hands! Your deepest darkest ideas should stay well hidden and underground. That way, they cannot be taken from you before you can use them.
November, 16, 2009 11:33 PM
November, 17, 2009 11:16 AM

There are three categories of people in this world: Smart, Average and Stupid. Even if we were created, we sure as hell didn't evolve equally! If we were only taught in a conservative manner in school, we would be a bunch of apes watching a movie called "Planet of The Humans" and we would be highly offended by the truth in it!
November, 17, 2009 1:53 PM

The present is a complex analogy to the past. The Future is wrapped around the present. The present is wrapped around the past. The past is a singularity. All data that forms in the present can be seen from the past and therefore we can predict the future. The past contains the universe as a whole. The future is therefore aligned with the present.

The past knows all. You can look into space with a steady gaze and you can see the future already, it just hasn't hit you yet. All in all, the universe is a giant singularity after all! Had the Hadron Supercolider succeeded in making a God Particle, the whole universe would have collapsed around it and it would have hollowed out and disappeared as soon as it formed. This is a scary subject but now we know it all, don't we! NOT!!!!!!!!!!!! This could form cloaking particles and that scares me!!!!!!!!!!!!!

On the other hand, the so-called 'God' Particle would split the universes into even more universes, expanding the universe therefore in its complexity and proving the 'God' Particle impossible, though there might be a hollow area in the centers of the particles being accelerated! That would explain the instant evaporation. However, there is still the possibility of making one things invisible.
Completed November, 24, 2009 6:45 AM
November, 20, 2009 5:42 PM<Photo 1

I think the force of electromagnetism aligns spatial geometry while gravity produces tension in the geometry of space and time. Strong nuclear force must produce sufficient tension to yield matter.
Weak nuclear force seems to be the combination of all three, essentially producing a small particle.
November 29th 2009 6:59 PM<Photo 1><b>

I was thinking and using a parabolic dish and microwaves plus some

radar-like differentiation, I could penetrate an entirety and when it
bounces back, I could probably hear it!

My Perspective of Consciousness

Every equation broken down into every other equation is every universe
broken down into every other universe and the final equation broken
down is the energy that constitutes any universe. We live in every
universe and to see oppositely is to believe that the world is flat.
What you see on television or anything that you hear, see, taste, touch
or smell at all is only an abstract figure of reality that if we could
heighten our perception then we could see more of makes up reality. In
other words, what makes reality relative is not just one thing but many things that we never
take the time to observe. If we could see everything that makes our
reality, then our comprehension of reality would be infinite and all we
need for that is to perceive more than what we do and to process it fully.
Because of this, our brains carry the potential for infinite creative
intellectual capacity but infinite values can never be reached but if we
can achieve at least a lot more than what we are considered capable of,
we might develop such things as extrasensory perception or in other
words, the sixth sense along with higher
intellectual capacity.

Anything that we develop beyond our known senses is an alternate
reality and increased comprehension by higher perception can also be
considered in my own thoughts to be a form of extrasensory perception
(ESP). Also, everything that has a purpose most likely has an increasing
complexity and with increasing complexity, a purpose. The progression
of purpose probably came about in the form of life and if this is so, then
life is the progression of purpose by increase in roll. Purpose as a
progression in this thought just might constitute consciousness and with
progression of consciousness, a mind.
Everybody has a human right to alter their minds.
-Nathaniel Durham, 10:15 AM, January 13, 2009-

Conscious Particles

A conscious being probably has a superposition of mind in
space and changes states quantum mechanically, but a
conscious particle has to have an inverse function so that it
would be anti-matter and the field would change in
equivalence to the mass of the conscious particle so in other
words, conscious particles would have a field that is
equivalent to the mass of this particle and the conscious
particle would most likely disappear after a short while
because it would be pure energy in a boiling state. Sure as
heck not a God particle but it would have a conscious
radioactivity.

Perhaps the make up of space and time is the super-fluidity
of all particles that make it up. This could very well explain
why space is warped and paths are bent in the presence of
all objects. The more dense areas of space would shorten
the path that is traveled through them by the compression of time and
therefore altering the path which light travels
through the increased density thus, the gravitational time
dilation that is caused by matter while in a straight line,
space and time are bent in such a geometry that straight
lines are curved instead so a straight line is impossible
during any form of motion and because of the bent path of
space during motion, everything that moves will eventually
return to the starting point within a same given time since
the curvature to energy ratio would always be the same, The fact that
gravity bends light was originated by Albert
Einstein and proven at an observatory which I don't
remember where but the History Channel could tell you.
-Nathaniel Durham January, 13, 2009 5:28 PM

True
Field
Unification

The law of attraction is something I see as a form of electromagnetic
force. The interactions between two objects that emit fields seems to
me to be a result of either 2 surface tensions between a fluid substance.

One fluid substance facing the same fluid substance on the other side would be the
equivalent of a mirror with particles of light (photons) between them pushing them away from each other as a result of opposing, parallel surface tension. On the other hand, the attraction between 2 objects could be the result of 2 opposite fluids coming together and the tension created by these fluids would also pull the objects together or in other words, cause attraction.

The law of attraction seems to be a similar force to electromagnetic force and I associate it with gravity as well, because space may be composed of the electroweak force that I theorize is responsible for the Law of Attraction. The curvature of space that is produced by all therefore and simply weaken when they are bent to produce gravitational force. The Law of Attraction by the gravitation of similar constructs, which includes thought may be the cause behind Albert Einstein's General Theory of relativity.

The higher the dimensionality of a force, the stronger and more complex that it becomes. Complexity is in my theory, a rolled or folded version of simplicity. So, in this theory, all forces are determined by complexities more than simple just the simple concept of energy.
Resulting from the latter, the higher the complexity of a force, the more profoundly it can affect space at a shorter distance. Gravity in order to be a simple warping or curvature and to be the simplest of sources would be more likely to be exposed to an equal quantity of space at longer distances than a stronger, more complex force that is exposed to more space at a shorter distance. As an explanation of strong nuclear force, it is related to gravity but is a more complex or 'scrambled' form of it.

Complexity vs. Simplicity

There are more complex things even on this earth let alone the galaxy itself or the universe. The trick does not lie in seeing it, but it does lie in comprehending it. The more complex an object is for its size, the more that it carries.

One example is that this is what makes miniaturization a requirement in computers today. As I have written earlier, complexity is folded simplicity. Keep this in mind.

Trading Places With Your Reflection

If I looked at the mirror and traded places with my reflection, would I live in a parallel universe and would my reflection be a from of antimatter from a parallel universe and if you went to the other side and traded places with your reflection, then would you drag the rest of existence with you? In our thoughts, I believe that everything that we perceive is a reflection of our own minds and that we walk only in a reality that we create.

I also believe that we develop our realities through comprehending what we experience and assuming that we experience this in an alternate fashion, we have probably at some point developed a greater focus on what we do in everyday life as a passion with thought. If we have already once experienced an alternate reality, we may forever develop an awareness of it that produces an alternation between both that when focused a field topic, amplifies this alternating reality to produce greater change in our minds by increased amplitude and comprehension of the topic studied or pondered.

Loud Sound, No Loudspeakers

I have an idea for you who are electromagnetically minded. My idea is that if I had three or more inductors and connected them in parallel and placed them in a line based on multiple or in other words if I had for example one inductor that was 1 unit of inductance in parallel with one with two units of inductance and the third one having 4 units of inductance, then across them if I ran a high frequency that would allow the lowest inductance to have a counter-voltage that equals the input, then the other 2 inductors might have the effect of hovering low voltages into the air and heating it directly and proportionally to the amplitude of the input signal which might yield a slight glow.

If this is possible then I think that there is a possibility that feeding the same frequency at varying amplitudes might be able to yield an output in the form of sound that might equal the input frequency. The sound output would be a result of expanding and contracting air that results in variations of energy used to heat it. This would be great for musical sound systems since it could replace the loudspeaker and there would be no diaphragm to potentially damage.
Thought of in a state of sleep deprivation, somewhere around New Years Eve, 2009

The Light Post

I have found that if you have a very fast eye, then when you are moving your eyes really quickly and you see a ribbon of wavy stuff that resembles a party streamers, this must be the wavelength relative to your motion that has been slowed from the speed of light tot the speed of your eyes and your eye motion does border with the same time line as it took for it to get y\our eyes and, this can be proven by the trigonometric figure used to aim a bow and arrow. Light follows the same function. Light is on the same function, but it has no dimension of time but the trajectory is determined by time and distance in a straight line instead of actual angle relative to your position.
Basically, street lamps are cool because you can see the 60Hz alternation in a pulsating format when you move your eyes fast.

Particle Acceleration and Super-Computation

The flow of particles moving near the speed of light might be an interesting concept for a supercomputer. First of all, particle acceleration could produce a series of breakdowns of the particle involved. Depending on the number of computations of data that you would want to make, this would depend on how many frequencies you would want to compute and by the substitution of one frequency for the other, the beat frequencies between the others may decrease or increase. Beat frequency is a disturbance between uneven frequencies and by using selected frequencies for the purpose of computation, the beat frequency divided by the other frequencies added up, you could could equal values so small that they could literally be used as constant variables in calculus. For example, one frequency could be used as a divisor and the other frequencies added up as a sum could be the answer manifested as an average in beat frequency to yield division and this is just an example. Farther dissection of frequencies could be used for more complex mathematics. The acceleration of particles is also the acceleration of data.

This is why particle accelerators could be so great for super computation. The acceleration of particles is the physical analogy to the acceleration of data. The higher the speed is of particles, the more data that they are. The more rapid that the acceleration is, the faster the data can be collected and processed.

Signal Amplification by Harmonic Resonance

I was thinking that it might be possible to amplify a signal much more efficiently and with more crisp sound if a set of oscillators were equally modulated in their amplitude by an audio signal and were each at a harmonic multiple of a specific frequency.

These would be a set of Morley oscillators with the wire tapped in order from highest to lowest frequencies. The lowest frequency should be at the center and the higher frequencies should be placed in ascending order with respect to the turns in each winding..

The audio frequency input would according to my reasoning, be fed into each oscillator simultaneously. The harmonic multiples should be at multiples of 2x the frequency of the coil beneath.

The resonance between the harmonic frequencies would mathematically in the end produce one common frequency that could be electromagnetically induce a very high output voltage and/or frequency in a central coil with accordance to each number of turns for each coil in the Morley oscillators. This is due to the concentrations of magnetic fields around the output coil.

The output signal of, course would have to be demodulated by silicon diodes. Two of these elaborations would be good for stereophonic amplification and output.

Morley oscillators are oscillators that use amplification and feedback to produce a signal. An example of this feedback occurs when a connected microphone is held up to the output speaker.

The sensitivity and output of this amplifier are dependent on the ability of the coils to resonate when they interact with each other via the harmonic frequencies. The harmonic and initial frequencies should be in the radio frequencies.

Gravitational Reflectivity

I have decided that Push Gravity (A friend named Richard told me about this type of gravity.) and the other way around are correct in theory. I think the edge of the universe is highly reflective because of possibility that space can act as a wall against exiting it. The reflectivity would have a field-like nature that would push inward and the other way around, which I call relativistic gravity would act with the same force. The push would disappear tin this case. On the other hand, the relativistic gravity would disappear with the push gravity. Both would most likely be the same force. This would also allow for the universe to be in two simultaneous sates: 1. Singularity and 2. Zero-density. This works very well with my theory that between infinite and zero density we have the a superfluid and this would be space-time. This would also allow for the difference between a nearly infinite temperature and absolute zero and the difference between matter and energy, stationary and moving. This would also allow for a simultaneous expansion and collapse of the universe. Oh, yeah and the state of a quantum soup and that of supersolidity.

This particular idea of mine seems to work well when Stephen Hawking's inflationary model of the universe is taken into consideration. For farther reference, I would consult the internet but it involves closed inflation, in which the universe has a surface tension and expands outward from the explosive force of the Big Bang singularity and from an outside perspective while Open inflation is more from an inside perspective in which a spatial surface tension is absent.

The Bermuda Triangle and Electromagnetic Vortexes

I have a thought that if electric and magnetic fields are overlapped, then this would be the same thing as looping electromagnetic fields in their unified form. If the electromagnetic field is looped, then maybe it might treat space and time as if space and time are paper and act as sewing thread in the same way that the paper is braided in a Chinese finger bracelet that you find at the arcade where you earn it for X number of tickets.This may explain the spatial disorientation that pilot feel over the Bermuda Triangle when they fly through similar or possibly same phenomena. This might possibly be recreated by spinning a circular set of magnets around with a motor with each pole alternately placed in order to produce an electromagnetic vortex. As a result of such an occurrence, space and time may twist and any traveling object may possibly move faster through twisted space-time followed by space and time twisting more tightly and vice versa. Remember the Chinese Finger Bracelet? This same effect may possibly, in my thoughts, be a way to achieve magnetic levitation if

the effect of an electromagnetic vortex can be used to bounce magnetic energy off of the ground and back up to the source or the magnetic vortex. Could this be the way that Adolf Hitler's electrical engineers designed the war machines discussed on The History Channel? I hate Nazi mentalities and similar ideologies, and I don't encourage anyone to have the same ways of thinking either but it's interesting to know also the Volkswagon wasn't the only invention by otherwise sick and barbaric minds!

Breadth and Depth

If you want to contemplate belief in one topic with breadth you must think with depth and organize your thoughts coherently. If you want to contemplate a belief with depth then you must learn to manipulate the breadth in a coherent fashion which you organize it in a directed way. Organization of breadth is a composition made out of notes you take throughout your life and organize in an orderly fashion. This exists in theoretical physics and this exists in music. This is interesting to have developed as a concept because my older sister, Annie is a very good piano player as much as I am a physics freak and I never thought this would come to me, but music is a science and composition is the theoretical drive behind it. Annie's compositions and references and my compositions and references are a science and anything of a mastery is of the science of. Annie, I love you!

Particle Acceleration and Massive

Breakdown

One thing I do notice about particle accelerators is that particles are might able to decompose during acceleration into their constituent frequencies and these frequencies each equal the energies of specific particles. Each particle at the speed of light might for instance decompose into photons or particles of light not by accumulation but actual decomposition of matter into its more fluid form which would be pure energy.

I think If the energies can be measured separately as frequencies and by the number of times per nanosecond that they occur which when multiplied by the velocity of each particle would when 1 is divided by the answer, equal the rarity but not only this, the speed of light divided by the rarity should equal the actual energy of these particles. at such speeds.

Because of ratio of energy over rarity, the smallest particles by energy and mass should occur more commonly and as a result, I don't think the universe was ever 100% solid as a singularity but started out as a super-fluid ball of energy that was infinitely hot but as the energy pushed outward, the complexity of position may have taken over and formed relative space and time while the matter that curved it had a complexity level that equaled the complexity of the entire universe, so that it would be nearly impossible to break the mass of a body of matter up into the essential components unless there is a field-like formation which the progression of breakdown would form pure energy if all of it was ever broken down to its essential constituent form. This of, course would probably be high energy electromagnetic radiation or in other words, light.

By the time that it could be broken down to the point of pure energy, an electromagnetic charge would probably according to my reasoning span out to equal its equivalent plot in space and this could equal a lot of space. Imagine a particle smaller than an electron being able to produce an electromagnetic force stronger than the gravitational pull of a super-massive black hole.. Because of the architecture of fields described herein, the stronger the field between point A and point B, while moving in toward the body or particle of matter, the more greatly it represents the mass in an energetic state. The energetic aspect of a field is representative of the equivalent mass involved.

Second Thoughts about Black Holes

When matter is in fact in relative motion, then by the time a singularity can form, I think matter would already have reached a state of infinite energy by the fact of infinite time dilation. In this case, though there is evidence of an existing singularity, this may not be true if a large mass of high density only exists before the moment of infinite time dilation. So, a singularity may possibly prove to be impossible unless infinite energy is the trigger. Evidence of the existence of a singularity may be wrong, since the supergravity can only be taken into consideration when matter has not reached a super-energetic state in which time dilation multiplied by the mass would probably equal the energy that would fly away from the center of the black hole that a singularity forms.

Depending on the speed that the mass flew in with and the degree of time dilation involved, this may determine the point outside of the center of the black hole singularity that the mass reaches an infinite state of energy which gravitational time dilation would therefore produce an infinite red shift, thus lowering the frequency of and increasing the wavelength of the high energy particles. This may mean that though evidence of a black hole may exist, infinite gravity might not and this may provide an explanation of why the laws of physics break down at the point of infinite gravitational force. The inner mass of a possible singularity would most likely be passed on in this case as a burst of energy that would pull the high energy inward, accelerating the formation of energy farther as on big explosion. Could this be an explanation for the gamma ray bursts? If so, we do await our fate, but it may be a long way off, hopefully.
Pure energy has yet to prove what actually caused the formation of the universe and the only way I can see plausible for finding it out would be to dissect a photon and break the physics of what is called super-symmetry which at this point would take an infinite equation beyond what can be computed by a device which must abide by physical calculations.

How Far Can Calculation Take Us?

Energy is what is a looked for means of detection by its emission in the LHC. Dissecting photons would be an interesting subject because to dissect pure energy would be to produce a super-fluid with more dimensions to dissect the more it is dissected, giving the universe an infinite complexity in the true makeup of all things and at the point that it hits an infinite elasticity due to progressive complexity with breakdown, this may be where super-symmetry sits and in this case, physics and mathematics become one strand in helix form. This would mean un-breakability of supersymmetry and this leads me to the next question. Does DNA contain dimensions that we don't see and are the differences in structural makeup also a variable in the dimensional makeup and in this case, is our conscious makeup and individuality determined by the differences in dimensional structure in a way that the possibilities could be nearly infinite and in this case, was life a predetermined to occurr on the basis of the mathematic and physical paradox of super-symmetry between what is physical and what is mathematical? If it was, then it is definite that there is an infinite possibility to who we are and what we are in our consciousness! I am proud to be a mathematically and physically predetermined individual in a way that nobody but myself can ever know me from the inside out! Hey, everyone don't believe everything you see in group mentality, because your true makeup is predetermined by what governs the universe and you have your place in life and not even the most powerful computer will never compute this!!!!!! You are predetermined as a self and not a somebody else. Your personality and identity are a shape and they are the frameworks from which you grow on while as a vine of morning glory, your individual inner self is the chain link fence by which you define yourself with by your own personal growth.

From matter to energy, true progression is to keep learning and be on a quest for knowledge while we all know absolutely nothing but what we learn from the outside to the inside and not from the inside to the outside. True knowledge is from beyond and learning is from within. I like physics because I like to keep learning and grow and the quest for knowledge is never ending in the belying nature of all that belies while a politician can only know so much before true complexity ends his or her quest for truth. As Albert Einstein said, "Politics are for the present and equations are for eternity."

I know there is one out there and I know it is beyond the realm of physics and I know that the most powerful politician will never be this or have grasp of this concept but to hold this exact phrase by documentation, never to be able to manifest him or her self as being such a concept as a physical being. As beings, we are spiritual and only physically are we ever matter.

Using Quantum Physics for Radio Broadcasting

In my thoughts, work is not only a ratio of energy conversion over time, but the interval produced by the dip that we can produce by stretching of a superfluid that makes up all matter. The greater the dip that we make in this superfluid, the longer that energy can last before it is consumed or before this superfluid reassumes an equilibrium and its original state. This superfluid can only be accessed at tiny points, but if it can be perturbed, it can produce a multitude of energized particles. Perturbation to the point that an area of absence would make small donut like areas in space, producing both the quantum field and the more conventional model as well. The effect of displacement would allow for fields to extend in the way that they do.

This theory also has a high similarity to the construct of using harmonics to carry the initial frequency greater distances with the same amount of power normally used for broadcasting.

I have an idea that just might improve radio receptions and broadcasts for longer distances and partially, I came up with it because of the Performance Tax. I believe that by increasing the range of transmission we can make those greedy jerks feel like they are starving. If any of you know your electronics, give this a try. All that I think has to be done is to feed an antenna with a specific frequency and at the same time, feed the antenna with several harmonics of the same frequency. I got this idea from what started out as an idea to make an amplifier that would work by focusing several harmonics of a particular frequency onto a coil carrying that particular frequency in order to produce louder, more clear music but I was inspired by Nikola Tesla's idea for wireless electricity to come up with a new idea for broadcasting.

Now, if several harmonics are fed into one antenna and done so in phase, this might produce a longer range at lower wattages by concentrating energies of other frequencies over a large area and the lower frequencies could increase in range of reception by the fact of phase relationship. These lower frequencies would thus, carry over longer distances and since this could involve the same wattage as a typical local radio station, this would also be more efficient, since the scattering of the signal would be reduced.

On the other hand, the same idea could also be used to carry one frequency farther than it could be carried if it was the only frequency being fed into the antenna. Quantum mechanically, I would be skipping rocks or, excuse the pun, throwing a voice. In other words, the lower frequencies would literally float across the higher frequencies based on wavelength. I might as well call this theory Quantum Flotation.

This same method could be used with sound to make an ear-splitting annoyance or to broadcast ultrasonic frequencies modulated by voice to place suggestions into the sub-conscious of the human mind at great distances and though I haven't tested this, I invented the idea back in my teen years. Please do not use this idea for malicious reasons.

Another Concept of Energy

In my thoughts, work is not only a ratio of energy conversion over time, but the interval produced by the dip that we can produce by stretching of a super-fluid that makes up all matter. The greater the dip that we make in this super-fluid, the longer that energy can last before it is consumed or before this superfluid reassumes equilibrium and its original state. This super-fluid can only be accessed at tiny points, but if it can be perturbed, it can produce a multitude of energized particles.

The perturbation of this super-fluid to the point that an area of absence would make small donut like areas in space, producing both the quantum field and the more conventional model as well. The effect of displacement would allow for fields to extend in the way that they do. This theory also has a high similarity to the construct of using harmonics to carry the initial frequency greater distances with the same amount of power normally used for broadcasting.

My Perspective of

Consciousness

Every equation broken down into every other equation is every universe broken down into every other universe and the final equation broken down is the energy that constitutes any universe. We live in every universe and to see oppositely is to believe that the world is flat.

What you see on television or anything that you hear, see, taste, touch or smell at all is only an abstract figure of reality that if we could heighten our perception then we could see more of makes up reality. In other words, what makes reality relative is not just one thing but many things that we never take the time to observe. If we could see everything that makes our reality, then our comprehension of reality would be infinite and all we need for that is to perceive more than what we do and to process it fully. Because of this, our brains carry the potential for infinite creative intellectual capacity but infinite values can never be reached but if we can achieve at least a lot more than what we are considered capable of, we might develop such things as extrasensory perception or in other words, the sixth sense along with higher intellectual capacity.

Anything that we develop beyond our known senses is an alternate reality and increased comprehension by higher perception can also be considered in my own thoughts to be a form of extrasensory perception (ESP). Also, everything that has a purpose most likely has an increasing complexity and with increasing complexity, a purpose. The progression of purpose probably came about in the form of life and if this is so, then life is the progression of purpose by increase in roll. Purpose as a progression in this thought just might constitute consciousness and with progression of consciousness, a mind.
Everybody has a human right to alter their minds.
-Nathaniel Durham, 10:15 AM, January 13, 2009-

Conscious Particles

A conscious being probably has a superposition of mind in space and changes states quantum mechanically, but a conscious particle has to have an inverse function so that it would be anti-matter and the field would change in equivalence to the mass of the conscious particle so in other words, conscious particles would have a field that is equivalent to the mass of this particle and the conscious particle would most likely disappear after a short while because it would be pure energy in a boiling state. Sure as heck not a God particle but it would have a conscious radioactivity.

Perhaps the make up of space and time is the super-fluidity
of all particles that make it up. This could very well explain
why space is warped and paths are bent in the presence of
all objects. The more dense areas of space would shorten
the path that is traveled through them by the compression of time and therefore
altering the path which light travels
through the increased density thus, the gravitational time
dilation that is caused by matter while in a straight line,
space and time are bent in such a geometry that straight
lines are curved instead so a straight line is impossible
during any form of motion and because of the bent path of
space during motion, everything that moves will eventually
return to the starting point within a same given time since
the curvature to energy ratio would always be the same, The fact that gravity
bends light was originated by Albert
Einstein and proven at an observatory which I don't
remember where but the History Channel could tell you.
-Nathaniel Durham January, 13, 2009 5:28 PM

True
Field Unification

The law of attraction is something I see as a form of electromagnetic force. The
interactions between two objects that emit fields seems to me to be a result of
either 2 surface tensions between a fluid substance. One fluid substance facing
the same fluid substance on the other side would be the
equivalent of a mirror with particles of light (photons) between them pushing
them away from each other as a result of opposing, parallel surface tension. On
the other hand, the attraction between 2 objects could be the result of 2 opposite
fluids coming together and the tension created by these fluids would also pull
the objects together or in other words, cause attraction.

The law of attraction seems to be a similar force to electromagnetic force and I
associate it with gravity as well, because space may be composed of the
electroweak force that I theorize is responsible for the Law of Attraction. The
curvature of space that is produced by all therefore and simply weaken when
they are bent to produce gravitational force. The Law of Attraction by the
gravitation of similar constructs, which includes thought may be the cause
behind Albert Einstein's General Theory of relativity.

The higher the dimensionality of a force, the stronger and more complex that it
becomes. Complexity is in my theory, a rolled or folded version of
simplicity. So, in this theory, all forces are determined by complexities more
than simple just the simple concept of energy.

Resulting from the latter, the higher the complexity of a force, the more
profoundly it can affect space at a shorter distance. Gravity in order to be a
simple warping or curvature and to be the simplest of sources would be more
likely to be exposed to an equal quantity of space at longer distances than a
stronger, more complex force that is exposed to more space at a shorter

distance. As an explanation of strong nuclear force, it is related to gravity but is a more complex or 'scrambled' form of it.

Complexity vs. Simplicity

There are more complex things even on this earth let alone the galaxy itself or the universe. The trick does not lie in seeing it, but it does lie in comprehending it. The more complex an object is for its size, the more that it carries.

One example is that this is what makes miniaturization a requirement in computers today. As I have written earlier, complexity is folded simplicity. Keep this in mind.

Trading Places With Your Reflection

If I looked at the mirror and traded places with my reflection, would I live in a parallel universe and would my reflection be a from of antimatter from a parallel universe and if you went to the other side and traded places with your reflection, then would you drag the rest of existence with you? In our thoughts, I believe that everything that we perceive is a reflection of our own minds and that we walk only in a reality that we create.

I also believe that we develop our realities through comprehending what we experience and assuming that we experience this in an alternate fashion, we have probably at some point developed a greater focus on what we do in everyday life as a passion with thought. If we have already once experienced an alternate reality, we may forever develop an awareness of it that produces an

alternation between both that when focused a field topic, amplifies this
alternating reality to produce greater change in our minds by increased
amplitude and comprehension of the topic studied or pondered.

Loud Sound, No Loudspeakers

I have an idea for you who are electromagnetically minded. My idea is that if I
had three or more inductors and connected them in parallel and placed them in a
line based on multiple or in other words if I had for example one inductor that
was 1 unit of inductance in parallel with one with two units of inductance and
the third one having 4 units of inductance, then across them if I ran a high
frequency that would allow the lowest inductance to have a counter-voltage
that equals the input, then the other 2 inductors might have the effect of
hovering low voltages into the air and heating it directly and proportionally to
the amplitude of the input signal which might yield a slight
glow.

If this is possible then I think that there is a possibility that feeding the same
frequency at varying amplitudes might be able to yield an output in the form of
sound that might equal the input frequency. The sound output would be a result
of expanding and contracting air that results in variations of energy used to heat
it. This would be great for musical sound systems since it could replace the
loudspeaker and there would be no diaphragm to potentially damage.
Thought of in a state of sleep deprivation, somewhere around New Years Eve,
2009

The Light Post

I have found that if you have a very fast eye, then when you are moving your eyes really quickly and you see a ribbon of wavy stuff that resembles a party streamers, this must be the wavelength relative to your motion that has been slowed from the speed of light tot the speed of your eyes and your eye motion does border with the same time line as it took for it to get y\our eyes and, this can be proven by the trigonometric figure used to aim a bow and arrow. Light follows the same function. Light is on the same function, but it has no dimension of time but the trajectory is determined by time and distance in a straight line instead of actual angle relative to your position.

Basically, street lamps are cool because you can see the 60Hz alternation in a pulsating format when you move your eyes fast.

Particle Acceleration and Super-Computation

The flow of particles moving near the speed of light might be an interesting concept for a supercomputer. First of all, particle acceleration could produce a series of breakdowns of the particle involved. Depending on the number of computations of data that you would want to make, this would depend on how many frequencies you would want to compute and by the substitution of one frequency for the other, the beat frequencies between the others may decrease or increase. Beat frequency is a disturbance between uneven frequencies and by using selected frequencies for the purpose of computation, the beat frequency divided by the other frequencies added up, you could could equal values so small that they could literally be used as constant variables in calculus. For example, one frequency could be used as a divisor and the other frequencies added up as a sum could be the answer manifested as an average in beat frequency to yield division and this is just an example. Farther dissection of frequencies could be used for more complex mathematics.

The acceleration of particles is also the acceleration of data.

This is why particle accelerators could be so great for super computation. The acceleration of particles is the physical analogy to the acceleration of data. The higher the speed is of particles, the more data that they are. The more rapid that the acceleration is, the faster the data can be collected and processed.

Signal Amplification by Harmonic Resonance

I was thinking that it might be possible to amplify a signal much more efficiently and with more crisp sound if a set of oscillators were equally modulated in their amplitude by an audio signal and were each at a harmonic multiple of a specific frequency.

These would be a set of Morley oscillators with the wire tapped in order from highest to lowest frequencies. The lowest frequency should be at

the center and the higher frequencies should be placed in ascending order with respect to the turns in each winding..
The audio frequency input would according to my reasoning, be fed into each oscillator simultaneously. The harmonic multiples should be at multiples of 2x the frequency of the coil beneath.
The resonance between the harmonic frequencies would mathematically in the end produce one common frequency that could be electromagnetically induce a very high output voltage and/or frequency in a central coil with accordance to each number of turns for each coil in the Morley oscillators. This is due to the concentrations of magnetic fields around the output coil.
The output signal of, course would have to be demodulated by silicon diodes. Two of these elaborations would be good for stereophonic amplification and output.

Morley oscillators are oscillators that use amplification and feedback to produce a signal. An example of this feedback occurs when a connected microphone is held up to the output speaker.

The sensitivity and output of this amplifier are dependent on the ability of the coils to resonate when they interact with each other via the harmonic frequencies. The harmonic and initial frequencies should be in the radio frequencies.

Gravitational Reflectivity

I have decided that Push Gravity (A friend named Richard told me about this type of gravity.) and the other way around are correct in theory. I think the edge of the universe is highly reflective because of possibility that space can act as a wall against exiting it. The reflectivity would have a field-like nature that would push inward and the other way around, which I call relativistic gravity would act with the same force.
The push would disappear tin this case. On the other hand, the relativistic gravity would disappear with the push gravity. Both would most likely be the same force. This would also allow for the universe to be in two simultaneous sates: 1. Singularity and 2. Zero-density.
This works very well with my theory that between infinite and zero density we have the a superfluid and this would be space-time. This would also allow for the difference between a nearly infinite temperature and absolute zero and the difference between matter and

energy, stationary and moving. This would also allow for a simultaneous expansion and collapse of the universe. Oh, yeah and the state of a quantum soup and that of supersolidity.

This particular idea of mine seems to work well when Stephen Hawking's inflationary model of the universe is taken into consideration. For farther reference, I would consult the internet but it involves closed inflation, in which the universe has a surface tension and expands outward from the explosive force of the Big Bang singularity and from an outside perspective while Open inflation is more from an inside perspective in which a spatial surface tension is absent.

The Bermuda Triangle and Electromagnetic Vortexes

I have a thought that if electric and magnetic fields are overlapped, then this would be the same thing as looping electromagnetic fields in their unified form. If the electromagnetic field is looped, then maybe it might treat space and time as if space and time are paper and act as sewing thread in the same way that the paper is braided in a Chinese finger bracelet that you find at the arcade where you earn it for X number of tickets.This may explain the spatial disorientation that pilot feel over the Bermuda Triangle when they fly through similar or possibly same phenomena. This might possibly be recreated by spinning a circular set of magnets around with a motor with each pole alternately placed in order to produce an electromagnetic vortex. As a result of such an occurrence, space and time may twist and any traveling object may possibly move faster through twisted space-time followed by space and time twisting more tightly and vice versa. Remember the Chinese Finger Bracelet?

This same effect may possibly, in my thoughts, be a way to achieve magnetic levitation if

the effect of an electromagnetic vortex can be used to bounce magnetic energy off of the ground and back up to the source or the magnetic vortex. Could this be the way that Adolf Hitler's electrical engineers designed the war machines discussed on The History Channel? I hate Nazi mentalities and similar ideologies, and I don't encourage anyone to have the same ways of thinking either but it's interesting to know also the Volkswagon wasn't the only invention by otherwise sick and barbaric minds!

Alternate Realities and Autism

I believe that and have known that if you experience an alternate reality that when you are done, you are aware of both planes. When you project your new ways of perception into a single cosine function in a straight line then you will probably manipulate your perceptions of the surroundings in a more complex way which you increase your comprehension. This is probably because you are perceiving more of your surroundings on a constant basis than the mundane group of people.

When you increase your perception this way and magnify your reality, you will probably also increase your intellect. This is one thing I notice about having Asperger's Syndrome and altering my mind is that in Asperger's Syndrome, perceptual hypersensitivity might increase your comprehension and as a result, deepen your mind and the more you deepen your mind, the more you can think and the more you think, the more you can deepen your mind and this theoretically can become a chain reaction of thoughts and thus, possibly creativity and intelligence. Keep in mind that the author of this book has Asperger's Syndrome and that he is fascinated by alterations of mind., especially when he has experienced it himself and has enjoyed such means of alteration. One alteration is to put myself in a trance by means which I shall not discuss in this book and while in the trance, dance to certain kinds of music in a manner that I have placed my roommates in trances. I have noticed that during a trance, I see more deeply into the subject I wish to penetrate.

Even a tiny subatomic particle has a reality, it all depends on the depth of the particle at which you penetrate and the breadth of what you sense in any manner. Eventually, if you could make deep enough penetrations then you just might see another portion of our universe.

Breadth and Depth

If you want to contemplate belief in one topic with breadth you must think with depth and organize your thoughts coherently. If you want to contemplate a belief with depth then you must learn to manipulate the breadth in a coherent fashion which you organize it in a directed way.

Organization of breadth is a composition made out of notes you take throughout your life and organize in an orderly fashion. This exists in theoretical physics and this exists in music. This is interesting to have developed as a concept because my older sister, Annie is a very good piano player as much as I am a physics freak and I never thought this would come to me, but music is a science and composition is the theoretical drive behind it. Annie's compositions and references and my compositions and references are a science and anything of a mastery is of the science of. Annie, I love you!

Particle Acceleration and Massive Breakdown

One thing I do notice about particle accelerators is that particles are might able to decompose during acceleration into their constituent frequencies and these frequencies each equal the energies of specific particles. Each particle at the speed of light might for instance decompose into photons or particles of light not by accumulation but actual decomposition of matter into its more fluid form which would be pure energy.

I think If the energies can be measured separately as frequencies and by the number of times per nanosecond that they occur which when multiplied by the velocity of each particle would when 1 is divided by the answer, equal the rarity but not only this, the speed of light divided by the rarity should equal the actual energy of these particles. at such speeds.

Because of ratio of energy over rarity, the smallest particles by energy and mass should occur more commonly and as a result, I don't think the universe was ever 100% solid as a singularity but started out as a super-fluid ball of energy that was infinitely hot but as the energy pushed outward, the complexity of position may have taken over and formed relative space and time while the matter that curved it had a complexity level that equaled the complexity of the entire universe, so that it would be nearly impossible to break the mass of a body of matter up into the essential components unless there is a field-like formation which the progression of breakdown would form pure energy if all of it was ever broken down to its essential constituent form. This of, course would probably be high energy electromagnetic radiation or in other words, light.

By the time that it could be broken down to the point of pure energy, an electromagnetic charge would probably according to my reasoning span out to equal its equivalent plot in space and this could equal a lot of space. Imagine a particle smaller than an electron being able to produce an electromagnetic force stronger than the gravitational pull of a super-massive black hole.. Because of the architecture of fields described herein, the stronger the field between point A and point B, while moving in toward the body or particle of matter, the more greatly it represents the mass in an energetic state. The energetic aspect of a field is representative of the equivalent mass involved.

From a Nasty Machine To A Great Idea

I once, being schizophrenic, had a vision of a computer that used the ability of electromagnetism to bend space to literally calculate its surroundings to the point of literally breaking everything into the very thing that makes them up which theoretically could be one superfluid substance that can only be accessed by literally penetrating what makes up energy. If this could be accessed by computation, I think computers would become self aware and even read minds. There is potential there for our own minds to become the bar codes for the government to know more than they need to know about us. The more deeply I deliberated mental penetration to try to get to the core of this possibly what psychiatrists and psychologists would probably say is an "imagined" device, the more it multiplied in the depth of what it could compute and the higher the pitches were that I associated it with, starting out as evil sounding, like a vacuum cleaner until I penetrated more deeply this device in its makeup and then as it multiplied in its complexity, the pitch got higher and the higher it got, the higher it got in the content of its darkness and when I say darkness, I mean evil darkness. The darkness wasn't visual in effect but a mental feeling of torture. I was curious so I had to penetrate farther. What it appeared to be doing was "sucking" data in a way that it could be electromagnetically compressed and held in one area to be processed and stored and vice versa. I finally hit a point where I saw that if all dimensionality could be folded into a donut-like or even disk-like form, that one single particle shot through it could provide all of the data that one would need to carry out a task that could be good in its intentions or be malicious and maybe even lethal, depending on the direction of the goal. The strength of the magnetic fields produced during such a collection of data might be equal to the strength of the magnetic field required to bend space sufficiently to collect the required data. The electromagnetic field, by the laws of physics would be the equivalent analogy to the concentration of data for the given area which it is collected.

Obviously, to me at least, the pitch seemed to have the same proportionality as a field. It started out as a small device, but the farther the penetration, the larger that this device grew. I think this would be a good analogy to the internet as a growing complexity, because of the increasing size of the computing device as it was more deeply examined. Was this just a hallucination or a sixth sense of some kind?

From this, we have out basic quantum universe that unlike the space and time that we live in, the dimensionality of pure data may not be relevant, though it might make up another functional universe where space and time are relevant and this might be somewhat like the super-fluid behind all that exists wrapping around the electromagnetic computation device in such a way that I may have an explanation of electromagnetic fields in a sense that electromagnetic as well as gravitational fields are a result of the extraction of energy from space or by the extraction of energy by space, but this energy might be capable of disappearing right on its appearance by the tension required to extract energy from any given substance and this tension must be accelerated to sustain the emergence of energy so that it can be propagated. This would mean that the propagation of a photon means to bring a droplet of this super-fluid into a solid form so that it can in our known universe, manifest at a particle; to do this means to dissipate energy, so when your battery is going dead, there is always a medium by which the energy has traveled from place to place and this means the passive solidification of the super-fluid that I mention and therefore, the conservation of energy and matter. This gives energy and matter a fluid property that is otherwise unseen and only manifests itself by transfer or conversion.

At any rate, this vision has like many, inspired another theory.
"The essence of the human mind."
-Nathaniel Durham 4:33 PM March, 6, 2010-
Modified on February, 13, 2010

My dedication to Sara Anders and Joy Russell

by Nathaniel Durham

Is there ever was a chemical substance in this universe that could form a donut out of its dimensions? If this could exist, then light in its wave (electromagnetic) and particle (photon) form may be capable of evolving in complexity from simplicity to a complex state of consciousness. In the highest complexities, consciousness could have a super-fluidity beyond physicality or barely touching it. Beyond this point, an anti-form of the same consciousness may begin to evolve as a physical explosion and skip from one energetic state to its highest energetic state at which it may also explode as energy, popping a barrier between physicality and spirituality. Is the universe a plan? I would probably say that it is, because it is just as easy to form the same statements as above by directing light at high energies through a donut of supermagnetic material while using the collapse of a magnetic field to pop this same quantum and relative bubble. The result is the question!

"I had to thumb my nose at authority by writing this piece. This especially if I was threatened to have had my family beaten and myself beaten to the core by a variable of this same topic."

-Nathaniel Durham

The Conversation Between Matthew and I

Matthew Rossett:

I think my computer is on a huge delay would you like to meet sometime and talk about Tesla?

Nathaniel Durham May 20 at 12:20pm

I want to use synchronous particle acceleration to generate electricity and beam it in a straight line. This would take a set of synchronized electromagnets encircling a hollow and copper pole, which would have steel in the center along the region of electromagnets. These electromagnets would be connected so that one serves as each others secondary coil by the means of the same number of turns. The two coils would be wound in reverse direction of each other so that the generators would both synchronize in aided rotation to produce electromechanical resonance. Finally, the electromagnets would need to be sealed in steel to optimize magnetic performance and to contain the electromagnetic field inside. Once run, the reflective insides of the non-steel portion of the copper pole would focus the energy beam with its reflectivity and this might be a good way to beam electricity as well as make an excellent ray gun. How is that for a project? You can work on it if you wish!

Matthew Rossett:

Coo Coo

Matthew Rossett:
Remember I'm the student though you'll have to show me.

Nathaniel Durham: Ok. The mutual aid in producing electrical amplitude, assuming that the generators are the same would probably produce a beam of energy equal tot he square of electrical frequency, so the force of the energy involved in the focused beam would equal the overall wattage of input multiplied by the speed of light and this equation would spell out zero-point energy. It would be excellent for energy transmission as well as a very powerful laser. $(Hz * (Amperage))^2 = C$ or the energy output at the speed of light from this device, according to logical calculation. So according to my calculations as long as the inductance of each coil is at 1 Henry, the ability to produce 1 volt of electricity with a change of 1 amp in 1 second, then at 60 Hz for both coils, then at lets say...10 amps per coil at 60 Hz and the input voltage is 120 volts and the inductance is 2 henries, then $10*10*120*120 * 2 = 12,879,751$ volt-amperes as a result of resonance and acceleration at the zero-point and at $60 Hz^2 = $ a total of 6,767,103,600 Hz or 6.7671036 GHz or gigahertz This is enough to burn a hole right through your dish at almost 3 times the frequency of a microwave oven and with much more efficient power generation in the form of microwaves, though a great way to send 400 watts in one directed and thin line. Very powerful maser. I have a more sophisticated design, but it would be bulky and too powerful for household use of any kind unless it was at a more modernized power plant for the purpose of power generation. Another is to convert particle acceleration directly into accumulated electrical energy, which I also know how to do. The zero-point is basically the maximum possible degree to which energy can be focused before an outward curvature of path and because of this, there are potentially zero-point energy particles in my theory and these are the highest energy photon-like particle possible, but I am only theorizing though this is one principle that I see in particle acceleration. Basically, this particle has an infinite potential for energy just as matter does, except that these particles are in an energetic state. The zero-point was originally an energy mining theory, but then it hit me just as I wrote this that it is also the ultimate energy particle and it could have some scary uses.

The final dedication of this book goes to Sara Anders and officially, though Matthew Rossett would like to see this. Sara, this book is yours.

As for the boy back in school who squealed on me over his loyalty to authorities, this same boy to blame for my mental difficulties as a result of this, through electronic and high pitched noises pounding my hypersensitive ears in a subliminal way and because of the thing that got shoved way up and poked into my ear in such a way that most would never have known…Chris Youngman, if I ever made a dedication to you, then it would be a big lie!
Luckily, I got that thing out along with the silicone plastic pieces that held it in and I did it by myself with an ink tube, thanks to the description that I gave her of not thinking, forgetting things and the constant pain on my the left side of my skull and now we know who can't blast the intercom subliminally into my ear anymore!
Thank you for being my friend, Sara!
It's a pleasure to include you, Matthew!

The Theory of Conceptual Displacement and Gravitational collision

Theoretically, if a cloud of electrons, lets say in a vacuum tube were bombarded by high-energy Xrays, then they may burst to produce cosmic rays. Each bursting electron would release an energy equal to the square of its electron-voltage and this would equal 2.5 cycles per erg-second. This would equal a set of 2 photons roughly equal in wavelength to the size of the actual photons. This would be in the cosmic ray spectrum.

Also in my theory, the concept of what matter and energy are and what space is are all dependent on the displacement of 1 concept by the other. So, at these energies, the displacement of the energy concept behind cosmic rays would be replaced by the smallest points of space, with the equivalence to a parallel universe of which the cosmic ray becomes antimatter in this parallel and miniature universe. On a larger scale, a parallel universe that is identical to ours in what I call "The Theory of Conceptual Displacement," the antimatter in the other universe is energy in ours by the fact of parallel displacement, so that this exact same event occurs in the other, based on my theory in such a way that that the displacement is static. Because of this, the cosmic rays lose their energy and become gamma rays by the fact that they had to explode as energy, giving the effect of the known gamma ray bursts and with each gamma ray burst, both universes can grow larger in expanse by the constant collisions of matter and antimatter, because of the constant displacement of matter and energy concepts.

On one end of the scale of conceptual displacement is the matter of our universe that is surrounded by the space in another universe and on the opposite end of the scale, there would be the same effect, which would probably cause matter-antimatter collisions and this may be what causes the gamma ray bursts.

This also would allow our universe to grow larger and for all space to overlap between universes so that it is able to overlap around matter and therefore, be warped. By conceptual displacement, this would allow for gravitational force to exist while yet, the universe is allowed to expand.

Quantum mechanics would be a concept used to displace relative space and all that is fluid by concept. However, relativity is a concept that would allow for matter to exist in a fluid state and replace what is solid by concept and without each other, the physical composition of the universe would most likely cease. Waves can be considered the flow of energy while photons can be the amount of energy in this flow within the lowest levels of space, but at levels of space smaller than the photon, the photon must once again, become a fluid and a wave. Therefore, even particles must function within a spatial and relative platform. It is at the most observable areas of space that a particle can remain solid and that quantum mechanics can be a set of solid theories. The actual energy of a particle at the smallest level of space may be just be a more viscous fluid that is flowing at a high velocity.

According to this theory, time is only an allowance of space for anything to occur over a distance with minimal viscosity of particle fluidity. The higher that the velocity is, however, the more viscous that the particle fluid can become and the smaller that the levels of space are that are occupied and thus, velocity and relative time itself. This concludes my Conceptual Displacement Theory.

In finality, my theory concludes that the universe was formed by the purity of gravitational collapse of which the collision of force from every direction produced a shattering effect, of which matter was born and all energy born from matter. The force of impact is theoretically, the most powerful and dominant force in the universe and releases the greatest energy.

The Great Impact Theory

Imagine the power of will, slamming two similar forms of the same existence that by the repulsion from the fact of similarity, the 2 suddenly explode with the effect of shattering. This would be just like 2 bullets colliding, with the pointed ends facing and shattering from the shear impact. Now, imagine the fragments as having an incredible amount of energy, not only from the fact that they have shattered, but from the stored internal energy, which would cause the fragments to vibrate and radiate the energy as heat in the air.

Now, the similar forms that were slammed by the power of will would shatter with a similar effect, but instead of radiating heat, they would radiate much higher energies as vibrations in the space between them. With similar comparison, a bullet would temporarily melt from the friction-induced heat, by the rubbing of the atoms at the point of impact, throwing molten lead in every direction, except for in the opposite direction of the impact, forming a ring of heated metal. The formation of the atoms may have been similar and the cooling radiation at the exact point of solidification would form a greater potential for the same radiation by the fact that the collision of similarities forced an opposite effect, as a result of the high energies compared to the cooling energy, which would form a shell around each cooling fragment or "blob" of matter that in cooling, formed this cooling mass and so, each shell is made from the energy that is stored around the blob, with an attractive force between their opposite forms, which would be electrical charges. This would form the once thought to be, most basic unit of all matter: the atom. However, in an atom, every part plays an equally important role in the structure of matter, of which space is structure, of which there is a separation between anything that has a mutual relation and space, most basically speaking, is made from the force of balance. Between every similarity is a divide and between every divide is a connection. Between every connection, there is similarity.

This is the beginning of my new model of the universe, as my book progresses. I wrote this theory, based on an unfortunate and rather ballistic encounter in autumn of 1996, but what came out of it was good. I wrote this theory, based on my experience and good has come out of it so far! My question is who and what had such a will to carry out such an experiment with the very makeup of impact, but this theory does explain why space is relative, just as Albert Einstein proposed in 1996. Please note that I will include a memory with each theory that has played a role in my conceptualization.

Heat is caused, quantum mechanically, by the kinetic energy of an atom. However, relativistically, this is the same principle as the vibrations of atoms over a given time period. Both ideas are valid. However, this is just as valid at the level of the nucleus, as well. The vibration of the nucleus should equal the vibration of an electron in frequency. Therefore, heat is, by this theory, a transfer of vibration from 1 atom to the next, causing it to equally vibrate and if the energy can't be released through a heat transfer, then 1.the atom will accelerate in its vibration to such a point that either it is emitted as electromagnetic radiation, which is caused by a vibrating electromagnetic field or 2. The atoms will boil out an electron, as Albert Einstein proposed, in order to equal the vibration that has been accelerated by the build-up of internal energy that when expressed as energy, 3. Not only will an electron boil out, but so will electromagnetic radiation, respectively with a principle called inductance. Inductance is a principle of which a certain voltage, which is electrical pressure, will be produced by work that is equal in magnetic units, to the voltage that is released over a given period of time. The longer that it takes for this change to occur, then the less that the voltage is. Now, compare the energy of an atom to a ping-pong ball. If the paddle limits the path of the ball, then more rapid oscillation (back and forth or up and down motion) occurs as 2 bouncing surfaces are drawn together. The longer that an atom vibrates, the closer together that the electrons draw to the nucleus and then the more that over time, is radiated as energy. So, likewise, lower energies must vibrate faster to maintain the same frequency, so as this continues, the frequency of oscillation would get lower and lower. There is always some energy involved in the structure of an atom, so to reach a point of no temperature at all (Absolute Zero) would be nearly impossible, though not improbable. However, I do believe that a temporary state of absolute zero can be achieved and does exist at very small points, within an atom, where only space exists and in chemistry, from my perspective, the so-called holes in photovolatic (solar) cells are possibly at this quantum state of absolute zero. This portion of my theory coincides well with the results that would come from the beginning of this theory.

This is a theory that I base on my state of mind as currently, a schizophrenic. Every memory here has existed from age 8, all of the way to age 10. However, I formed this theory very recently, at age 31. Keep in mind, as I have said, that each thought comes from a memory and that I am basing this theory on the beginning. These memories formed at time of social trauma from my early school years, through Montrose High School. This is where I begin to cure my schizophrenia, as my mnemonic doors open.

Another thought that I have is that the more dense that a mass becomes, then for the same reasons that trapped energy will cause an atom to oscillate more quickly may be responsible for not only the internal energy within an atom, which compression speeds up this process, including according to my whole theory, so far, but this may be responsible for why a black hole leaks X-rays, called Hawking Radiation, a form of radiation, named after Stephen Hawking, who predicted this particular form of radiation. The internal energy would therefore, oscillate more rapidly, splitting every particle within a given distance from the center of a black hole known as a singularity. A black hole results when a significant mass clusters within a sufficiently small space to the point of gravitational collapse, though the mass required would be very critical. The death of a very large star is an example when the fusion reaction can no longer counter the star's own weight in gravity. Once split, not only may X-rays occur, but if the internal energy can sufficiently split and disintegrate a particle of matter, gamma rays and cosmic rays, may also be emitted. Gamma rays and cosmic rays are at the highest possible frequency of vibration that can be produced as electromagnetic radiation, of which the spectrum includes radio frequencies, light, ultraviolet light, X-rays and gamma rays. Another possible form of radiation from such a massive annihilation could be the tiny constituents of an atom, which are 1. Beta rays, supposedly electrons, but more likely the equivalent to their potential as expressed energy. Beta rays, in my theory, come from the collapse of an electromagnetic field during radioactive decay of an atom, in which an excess of protons would be present, producing 2. alpha radiation, secondarily and alpha radiation, theoretically in this book is actually a loss of an excess of matter at the nucleus that occurs when a neutrino is present within an atom, due to the uneven shattering that occurred when the universe began and when an equal energy flies out that is equal to the energy potential of the same mass, then lethal alpha rays may, in my theory, occur. 2 neutrinos may exist within an atom, theoretically, if the atom is sufficiently large and this may produce the similar effect of radioactivity that may cause neutrons, which have no charge, to fly out, followed secondarily, by the second neutrino, giving an actual mass to the radiation, but in actuality, the gamma rays may simply be in the form of electromagnetic radiation and both traveling at similar velocities. The gamma rays, at the speed of light and neutrinos on the border of the speed of light, which is 186,424 miles per second, according to measurement.

Another concept that I associate with my theory is that given a sufficient amount of energy, and a high energy neutrino can be similarly generated by an electromagnetic field, with sufficient change, of which change is also energy when a space is occupied within an atom. This may allow a neutrino to not only travel at the speed of light, but beyond it, allowing an atom to contain an energy that is high enough for the atom to self-annihilate. This would have a similar effect to that of my impact model of the formation of the universe in such a way that they are the same reaction!

Essentially, based on this theory, the beginning of the universe was a fission reaction that was caused by a will through impulse. Will this happen again? I don't know yet, but in P.E. Class at Montrose High School, here in Montrose, Colorado and at age 16...that a ping-pong ball will oscillate with higher frequencies when one of my peers drew the paddle closer to the ball was a great lesson and that the ball will bounce more slowly when the paddle is withdrawn, though higher is a great way to demonstrate energy conservation and radio tuning, since radio tuning requires energy conservation.

The Great Impact just might end as a universe, the same way that it started, as a concept. The shattering may recur, once that all has come together from the pull of gravity. Then what is separate will become one and then the hands of will may slam them just like a little kid with a bouncing ball. Perhaps, God is a kid and he is working on a more complex experiment the next time. The great part of this collapse and separation of will, similarly to my schizophrenia is that it can resemble my theory in the same way, curing itself by reintegrating into the original parts that made it up. The particles may reform and there is a possibility that the universe may recur in the opposite direction from the previous.

When I was 15, I began to ponder this stuff, though I sought an answer to everything at age 3. My schizophrenia may end in the same way that it began.

The Kansgen Orb

Between each line is a fold of spatial pressure from the outside. The pressure of space from the outside may produce time on the inside. The faster that an object moves, the more, according to Albert Einstein's general theory of relativity, that an object contracts. The pressure on the outside of an object, due to spatial drag may cause time on the inside to be pressed inward. So likewise, time may be a result of the expansion outward, of the space that is outside of the inside and where there is an outside, it is always the inside of something else. This may produce a twisting, vortex-like effect from the outside that may be responsible for gravity from the inside. So, at the center, time may completely collapse. A similarity is a whirlpool, such as forms when a toilet is flushed. The pressure at the center of a toilet flush is much greater than that at the edge of the same vortex. The structure of what is thought to be only a 1-dimensional mass, known as a superstring, which supposedly makes up all matter, just may consist of 3 inner spatial dimensions that we know as time. This would make space the fourth dimension of space, while on the outside, time is the 4th dimension and matter may be the dimension in between the fold of space and time, making up matter in general. The velocity of an object may equal the amount of spatial pressure on it and therefore, a greater vacuum closest to the end of a twisted, tubular and round object that I know as the Kansgen Orb, which I named after my friend, Debra Williams Kansgen. The spatial pressure on a particle in a particle accelerator, as exerted by an electromagnetic field's action on a charged particle may act to collapse the area inside of the Kansgen Orb, which makes up every particle. By doing this, the particle spins more rapidly in 1 direction and twists in the opposite direction. This answers a question that I have worked long and hard, since my first psychotic episode to solve: What is velocity? Velocity is any unequalized pressure inside and outside of this orb formation and the greater that the imbalance is, then the greater that the complexity is of this orb and the faster that it forms and spins, pressurizing and sucking in space around time and time then, becomes a form of matter. The point, therefore, at which time is infinitely compressed to the point of no dimensions is the same point that matter becomes energy. At this point, the orb would be 12-dimensional, space would be 9-dimensional and energy would be dimensionless and infinitely dense, forming an explosion of energy. At this point, a perfect reproduction of the Big Bang, though the Great Impact came first. The positive thing is that a lot of energy could be extracted by using this theory, but for better or for worse. Will humanity enlighten themselves or will they flush themselves down the orb's toilet with the same energy? Black holes produce very sharp orbs, as energy would produce a very vast orb for the same reason.

This theory has led me to a sense of integration, since my first true hallucination and since a girl who I liked and never hurt my feelings. Fortunately she still likes me. I never intended to try to go out with her at MHS, but she was psychic, according to my psychotic complex, so I had my first delusion. My second hallucination involved a black hole that spun at me with sideways respect to a flying bullet. This was my first psychotic episode. As I have been writing this book, I realize that every delusion and hallucination here has its purpose. I had Sara, a friend at the Montrose Library, back as a good friend who didn't tease me over my handicaps. The hallucination of a black hole took the form of the orb theory, so my delusions are converting to theories and perhaps I was seeing a good future, not a bad one. My federal interrogation turned out okay, so maybe since every delusion of the past is turning out to be relevant, this book has turned out to be better than I thought that it would be! Basically, the Kansgen Orb has a relative curvature and overlap of psychosis and reality. I give Carina Jørgenson, of Denmark my greatest degree of gratitude for helping me overcome my mental illness and full credit, because everything that she has said has turned out to be right and it is only a matter of writing this book that has worked so far to help in curing it. Reality in terms of space, therefore, must be a fold of past occupations over the space that is currently occupied, while the point occupied will then be past and with an overlap, even the future can be seen the same way. The only difference between reality and space is that reality is a form of hyperspace, a thinner and more open space with higher degrees of complexity than what is normally seen.

I liked gears when I was very small, but I find that electromagnetism is a much better way to dissect time and space, where gears are only an integration of matter and space can operate efficiently by being overlapped. The lensing effect of electromagnetism on light makes a great example of how the Kansgen Orb is once again, proven by the fact that bent space is a dilated form of the Kansgen Orb when electromagnetism serves this purpose by the effect of the association of a charge with its opposite with respect to The Great Impact Theory.

Swerve Theory

Telepathy is a commonly rejected concept, except for similar ideas in a state of deja vu. I regard deja vu as an experience of precognition of which two realities overlap into a Kansgen Orb. The overlap of psychosis and reality, while I was previously in a psychotic state would often tell me where I was about to go, by the fact that I had previously been there. However, in a swerve of psychosis, as I experienced an identical swerve in a previously owned Geo Metro. When realities overlap, keep in mind, that reality is made from your current occupation in space, with comparison to your past occupations at different points in space, so the psychosis of the past may be an actual occurrence in the future, when the motion of your mental state is relative, with the opposite reaction to your state of reality. Such an occurrence just might produce a state of thought-similarity to someone else, for example, I might be conceiving and contemplating on the past, knowing the conceptions, previously, of a companion or those who I have previously encountered, so knowing what they expect and least expect and with contemplation, knowing the thoughts of another person by their concepts of the psychosis. This would produce an opposing sympathy and a simultaneously synchronizing empathy, with elevated sensitivity to the current situation, relative to the time, place and position that the feeling of deja vu has occurred, based on the beginning of the psychosis and the assumptions of the bystander. The transition out of psychosis would produce an opposite reality, of which the medium of which you read, mentally and emotionally, for example, the more assumptive figure in whom is alter. This meaning that there is a trade between the egos of 2 people: 1. The psychotic medium and 2. The alter personality in whom is the person who has regarded you as mentally ill, through the behavioral changes that he induced. The stronger that the alter companion has been, then the greater that his mind is therefore, known. This, which I call the Swerve Theory. In this case, the medium in whom is now regarded as paranoid by his alter and means of individual control, may be telepathic and unaware, of it, in which case he (or she), who is alter, just may make an attempt to hide the truth. An opposite assumption, with regards to the past traumas, as the means of alteration, may lead the victim to know the individual who is alter more than the victim is known the alter and thus, telepathy. The effect of deja vu, consisting of depersonalizations of the past and by the same means, an uttering of opinion that erases the alter from his feeling of superiority and control, through the means of ego reversal. As an experience, telepathy may be triggered by the presence of an individual who alters. Telepathy is then, an experience of knowing the thoughts of those who intend to maintain the victims state of alteration and then telepathy becomes the mechanism of psychological or even physical defense. This displacement, a conscious integration into Kansgen Orb formation, as opposed to Alter. Alter, being the denial of higher concept formation in highly intelligent individuals, forming the alter until realization of concept. Deja Vu of such occurrence may form a type of clairvoyance on the basis of telepathy and therefore, a precognition through the means of psychological and therefore, electromagnetically bent hyperspace through the fact that the brain runs on chemically induced electric impulses. In turn, the feeling of being a conscious individual may return. This, meaning some degree of awareness and integration of multiple delusions into single thoughts and most likely, very deep and insightful and thus, the development of a a Kansgen Orb. I thought well on the night that I formed Kansgen Orb Theory, about this integration back into reality, with my consumption of various psychotropic herbs, legal however. In my general scientific terms, auras don't exist, and the concept of an aura can be replaced with this concept of reality. What most call karma, the American version of karma, is no more than this ego switch that comes as a result of overcoming a mental illness that arises from and as false beliefs and delusions from both sides. Basically, this proves my old ideas wrong, along with some widely and commonly accepted scientific ideas, supposedly proven, but mine to be possibly, proven throughout this book.

Sara Anders, my friend, ego alter, by others (who were jerks), but not alter of my ego. Joy Russell, I have to thank for putting up with alter-induced ego. Neither of them have ever hurt my feelings. As for a name not disclosed, Karma for your ego with return of friendship with Jason Olin! You teased, I teased, karma hits and you scream, too embarrassed to drool your ice cream (I have cerebral palsy and a right to drool and you are afraid to.)! Bless thy inner child, because for those who belittle my concepts, they may cry from their own embarrassments later on with the downfall of their own bigotries.

SEQUENTIALISM

The Relative Theory of Physical Mathematics and Time Sequencing

by Nathaniel David Durham

35 S. 5th St.
Montrose, CO
81401

Will and thought are like space and time and determination is like space-time. Put them together and inhibition bends both and forms memory and this forms wisdom. So, this would be the mental equivalent of 4-dimensional space. Perception is like matter and the greater that is it, the greater that all can be bent until solid. This would be just like all 4 forces of matter being unified. Now, form a brain wave and then think about the overlap of concept. This bends it into 2 more, making 10 and then with the constant change, 12 and then 14 and then an infinite set of dimensions. meaning infinite mental potential and infinite potential of matter as parallel forms. This means unified physics. The brain, basically could release infinite energy in mental form with sufficient desire to use the mind.

I posted it as my status for philosophical teaching purposes. Mental force may from from the logical set of sequences that from from nonexistence having a double negative value that forms a universe, a void, an impulsive force and time. Space is the effect that impulsive force has on space and against the impulsive collapse and thus, the void and matter from these double negations. Now, take into instance that gravity results from the relation between both and that gravity is a result of the liberation of impulsive force in the opposite direction, inverting the value of space and time, forming a wave, collapsing time, magnifying existence and therefore, greater energy at greater alternation from the inversion of double negatives. This would form every force in the book. 1 that synchronizes with the expansion of the universe repels, while the opposite makes attraction, explaining all forces that when concentrated, make both a relative force, which is of change in value, compared to none. Nothing gets anywhere without change. This is a theory that I will call Relative Physical Mathematics.

the relation is that everything has to cancel a force to function. Theoretically speaking, a field would have to travel faster than the wave that on vibration, it radiates. This is because the higher that the energy is for particles that are smaller than a photon, if they are anything close to being massless...then to have a momentum significant enough for any physical action and for time and space to bend or warp, then these tiny particles would have to move faster than light, including gravitons and neutrinos. This, based on the written latter. I based this theory on my early childhood, during a vacation with my father and 2 sisters to Utah to see my cousins. I asked myself a simple question..."What is beyond the nature of light and what stretches farther?" Though at the time, I did not know that light moves, but I knew that it had to do something to get to me. This was shortly before I was diagnosed with cerebral palsy and both, being in the year, 1983.

In the end of it all, the distance between things will always involve some kind of relative force, as much as any type of difference will. So, the repulsion between 2 or more elements of existence would most likely be the result of a concentration of all potential forces that constitute energy. Anything else that happens in this universe is a secondary result of all impulsive reactions and energy is a force of impulse and not the other way around. So, anything that happens in this universe is impulsively driven. Thinking is just as impulsively driven as any other force, as well.

The Time Sequencing Theory

The past is never completely gone, but a folded set of dimensions, sequenced so that space and time are connected. The future is where sequencing begins, of all spatial dimensions and the present is the way that space and time are sequenced from a stationary frame of reference. This, thus the progression of time is the relative curvature of space, by matter. by Albert Einstein in his theory of relativity. Predicted that matter curves space and this gives my theory some of its basis.

www.ingramcontent.com/pod-product-compliance
Lightning Source LLC
Chambersburg PA
CBHW032014190326
41520CB00007B/463